Vitamin C: Current Concepts in Human Physiology

Vitamin C: Current Concepts in Human Physiology

Special Issue Editors

Ramesh Natarajan
Anitra C. Carr

MDPI • Basel • Beijing • Wuhan • Barcelona • Belgrade

MDPI

Special Issue Editors
Ramesh Natarajan
Virginia Commonwealth University
USA

Anitra C. Carr
University of Otago
New Zealand

Editorial Office
MDPI
St. Alban-Anlage 66
Basel, Switzerland

This is a reprint of articles from the Special Issue published online in the open access journal *Antioxidants* (ISSN 2076-3921) from 2017 to 2018 (available at: https://www.mdpi.com/journal/antioxidants/special_issues/vitamin_C)

For citation purposes, cite each article independently as indicated on the article page online and as indicated below:

LastName, A.A.; LastName, B.B.; LastName, C.C. Article Title. *Journal Name* **Year**, *Article Number*, Page Range.

ISBN 978-3-03897-294-5 (Pbk)
ISBN 978-3-03897-295-2 (PDF)

Contents

About the Special Issue Editors vii

Preface to "Vitamin C: Current Concepts in Human Physiology" ix

Gina Nauman, Javaughn Corey Gray, Rose Parkinson, Mark Levine and Channing J. Paller
Systematic Review of Intravenous Ascorbate in Cancer Clinical Trials
Reprinted from: *Antioxidants* **2018**, 7, 89, doi: 10.3390/antiox7070089 1

Mike N. Foster, Anitra C. Carr, Alina Antony, Selene Peng and Mike G. Fitzpatrick
Intravenous Vitamin C Administration Improved Blood Cell Counts and Health-Related
Quality of Life of Patient with History of Relapsed Acute Myeloid Leukaemia
Reprinted from: *Antioxidants* **2018**, 7, 92, doi: 10.3390/antiox7070092 23

Nicola Baillie, Anitra C. Carr and Selene Peng
The Use of Intravenous Vitamin C as a Supportive Therapy for a Patient with
Glioblastoma Multiforme
Reprinted from: *Antioxidants* **2018**, 7, 115, doi: 10.3390/antiox7090115 30

Maria Leticia Castro, Georgia M. Carson, Melanie J. McConnell and Patries M. Herst
High Dose Ascorbate Causes Both Genotoxic and Metabolic Stress in Glioma Cells
Reprinted from: *Antioxidants* **2017**, 6, 58, doi: 10.3390/antiox6030058 36

Melissa Prier, Anitra C. Carr and Nicola Baillie
No Reported Renal Stones with Intravenous Vitamin C Administration: A Prospective Case
Series Study
Reprinted from: *Antioxidants* **2018**, 7, 68, doi: 10.3390/antiox7050068 49

Anitra C. Carr
Symposium on Vitamin C, 15th September 2017; Part of the Linus Pauling Institute's 9th
International Conference on Diet and Optimum Health
Reprinted from: *Antioxidants* **2017**, 6, 94, doi: 10.3390/antiox6040094 59

Karel Tyml
Vitamin C and Microvascular Dysfunction in Systemic Inflammation
Reprinted from: *Antioxidants* **2017**, 6, 49, doi: 10.3390/antiox6030049 68

**Gwendolyn N.Y. van Gorkom, Roel G.J. Klein Wolterink, Catharina H.M.J. Van Elssen,
Lotte Wieten, Wilfred T.V. Germeraad and Gerard M.J. Bos**
Influence of Vitamin C on Lymphocytes: An Overview
Reprinted from: *Antioxidants* **2018**, 7, 41, doi: 10.3390/antiox7030041 79

**Kimberly Sanford, Bernard J. Fisher, Evan Fowler, Alpha A. Fowler III and
Ramesh Natarajan**
Attenuation of Red Blood Cell Storage Lesions with Vitamin C
Reprinted from: *Antioxidants* **2017**, 6, 55, doi: 10.3390/antiox6030055 93

Juliet M. Pullar, Simone Bayer and Anitra C. Carr
Appropriate Handling, Processing and Analysis of Blood Samples Is Essential to Avoid
Oxidation of Vitamin C to Dehydroascorbic Acid
Reprinted from: *Antioxidants* **2018**, 7, 29, doi: 10.3390/antiox7020029 106

Juliet M. Pullar, Anitra C. Carr, Stephanie M. Bozonet and Margreet C. M. Vissers
High Vitamin C Status Is Associated with Elevated Mood in Male Tertiary Students
Reprinted from: *Antioxidants* **2018**, *7*, 91, doi: 10.3390/antiox7070091 . . . . **117**

Stine Normann Hansen, Anne Marie V. Schou-Pedersen, Jens Lykkesfeldt and Pernille Tveden-Nyborg
Spatial Memory Dysfunction Induced by Vitamin C Deficiency Is Associated with Changes in
Monoaminergic Neurotransmitters and Aberrant Synapse Formation
Reprinted from: *Antioxidants* **2018**, *7*, 82, doi: 10.3390/antiox7070082 . . . . **126**

About the Special Issue Editors

Ramesh Natarajan, PhD, was Professor of Medicine at Virginia Commonwealth University, Richmond, VA, USA. Dr. Natarajan's research focused on pulmonary/critical care medicine and in particular on sepsis, thrombosis and hemostasis, trauma, hemorrhagic shock, resuscitation, brain injury, and vitamin C. He conceived the notion of treating sepsis with parenteral vitamin C and designed animal models of sepsis to test his hypothesis. He subsequently participated in a Phase I and Phase II study of intravenous vitamin C in sepsis. He has published 69 manuscripts, 4 textbook chapters, has given over 130 presentations, and has published 140 abstracts. He serves on several Editorial Boards, as an ad hoc reviewer for journals and study sections for NIH, CDMRP, and VA. He is currently working on Combat Trauma Research.

Anitra C. Carr, Associate Professor, is a Sir Charles Hercus Health Research Fellow and Principal Investigator at the University of Otago, Christchurch, New Zealand. Dr Carr is particularly interested in the role of micronutrients in human health and disease and has spent much of her research career investigating the antioxidant and health effects of vitamin C. In 1998–2001, she carried out an American Heart Association Postdoctoral Fellowship at the Linus Pauling Institute, Oregon State University, Corvallis, USA, under the mentorship of eminent vitamin C researcher, Prof. Balz Frei. Whilst there, she produced a number of high impact publications on vitamin C in human health and disease. Recently, Dr Carr has been carrying out human intervention studies investigating the bioavailability and health effects of both oral and intravenous vitamin C, particularly its roles in acute and chronic diseases, such as cancer and infection, subjective mood, and quality of life.

Preface to "Vitamin C: Current Concepts in Human Physiology"

Vitamin C is synthesized by almost all animals. However, for humans, it is a vitamin that needs constant replenishment in the diet. While its role as an anti-oxidant and for preventing scurvy have been known for a long time, novel functions and unrecognized associations continue to be identified for this enigmatic molecule. In the past decade, new details have emerged regarding differences in its uptake by oral and intravenous modes. While vitamin C deficiency remains largely unknown and poorly addressed in many segments of the population, novel pharmacological roles for high-dose, intravenous vitamin C in many disease states have now been postulated and investigated. This has shifted its role in health and disease from the long-perceived notion as merely a vitamin and an anti-oxidant to a pleiotropic molecule with a broad anti-inflammatory, epigenetic, and anti-cancer profile.

This Special Issue comprises original research papers and reviews on vitamin C metabolism and function that relate to the following topics: understanding its role in the modulation of inflammation and immunity, therapeutic applications and safety of pharmacological ascorbate in disease, and the emerging role of vitamin C as a pleiotropic modulator of critical care illness and cancer.

Ramesh Natarajan, Anitra C. Carr
Special Issue Editors

antioxidants

MDPI

Review

Systematic Review of Intravenous Ascorbate in Cancer Clinical Trials

Gina Nauman [1], Javaughn Corey Gray [2], Rose Parkinson [2], Mark Levine [1] and Channing J. Paller [2,*]

[1]   National Institutes of Health, National Institute of Diabetes and Digestive and Kidney Diseases,
      Clinical Nutrition Section, Bethesda, MD 20892, USA; gn157@georgetown.edu (G.N.);
      markl@bdg8.niddk.nih.gov (M.L.)
[2]   Sidney Kimmel Comprehensive Cancer Center, Johns Hopkins Medical Institutions, Baltimore,
      MD 21287, USA; corey.gray@jhmi.edu (J.C.G.); rparkin1@jhmi.edu (R.P.)
*   Correspondence: cpaller1@jhmi.edu; Tel.: +1-410-955-8239; Fax: +410-955-8587

Received: 10 May 2018; Accepted: 6 July 2018; Published: 12 July 2018

Abstract: Background: Ascorbate (vitamin C) has been evaluated as a potential treatment for cancer as an independent agent and in combination with standard chemotherapies. This review assesses the evidence for safety and clinical effectiveness of intravenous (IV) ascorbate in treating various types of cancer. Methods: Single arm and randomized Phase I/II trials were included in this review. The PubMed, MEDLINE, and Cochrane databases were searched. Results were screened by three of the authors (GN, RP, and CJP) to determine if they met inclusion criteria, and then summarized using a narrative approach. Results: A total of 23 trials involving 385 patients met the inclusion criteria. Only one trial, in ovarian cancer, randomized patients to receive vitamin C or standard of care (chemotherapy). That trial reported an 8.75 month increase in progression-free survival (PFS) and an improved trend in overall survival (OS) in the vitamin C treated arm. Conclusion: Overall, vitamin C has been shown to be safe in nearly all patient populations, alone and in combination with chemotherapies. The promising results support the need for randomized placebo-controlled trials such as the ongoing placebo-controlled trials of vitamin C and chemotherapy in prostate cancer.

Keywords: intravenous; ascorbate; vitamin C; clinical trials; cancer; patients

1. Introduction

Ascorbate (vitamin C) was proposed to have anticancer effects as early as the 1950s [1,2] However, the earliest effort to using high-dose vitamin C—both intravenously (IV) and orally—as a cancer treatment occurred in the 1970s, by Scottish surgeon Ewan Cameron and his colleague Allan Campbell. For comparison purposes, in 1974, the recommended dietary allowance of vitamin C was 0.045 g (45 mg) daily [3]. Cameron and Campbell treated 50 patients with various types of advanced cancers with high doses of oral ascorbate, IV ascorbate, or both. Several responses were observed following this treatment [4–6]. These findings led to a collaboration between Cameron and Nobel Prize winning chemist Linus Pauling on the evaluation of two case series of cancer patients [5,7]. The data obtained from these cancer patients suggested that there was a potential survival benefit when their treatment was supplemented with oral and IV vitamin C [7,8]. Limitations of these findings have subsequently been described [9], including that the findings were retrospective, without controls or blinding, and that studied patients may have been at risk for endemic vitamin C deficiency.

To test ascorbate prospectively, two randomized, placebo-controlled prospective trials were conducted at the Mayo Clinic, in which cancer patients received either placebo or 10 g of oral ascorbate. Each study noted no significant difference between the ascorbate-treated and placebo-treated groups [2,10]. Based on these results, ascorbates role in cancer treatment was dismissed [11,12].

However, there was renewed interest in the use of vitamin C as a cancer treatment, based on the discovery that intravenous ascorbate produced plasma ascorbate concentrations that were much higher than those from oral ascorbate, and were not possible from oral ascorbate [9,13,14]. Although Cameron's subjects received both intravenous and oral ascorbate, subjects in the two randomized placebo-controlled trials at Mayo Clinic received only oral ascorbate. The significance of this key difference was not previously recognized until ascorbate pharmacokinetic studies in healthy subjects revealed the importance of the route of administration.

Subsequently, emerging preclinical and clinical studies led to a revival of interest into the clinical potential of intravenous ascorbate as a cancer chemotherapeutic agent, specifically its synergy with chemotherapy and amelioration of chemotherapy-induced side effects [15]. Additional studies on the efficacy of vitamin C as a therapeutic have shown that intravenous administration achieves high plasma concentrations that are not achievable through oral administration [13,16–18]. Specifically, oral administration of vitamin C at a dose of 1.25 g achieved a maximum plasma concentration of 134.8 ± 20.6 μmol/L (μM), while IV administration of vitamin C achieved a maximum plasma concentration of 885 ± 201.2 μmol/L [13,16]. In the text that follows, we refer to plasma ascorbate concentrations as pharmacologic when they can only be achieved by intravenous administration in humans, and as parenteral (intravenous or intraperitoneal) administration in rodents.

The role of intravenous vitamin C in combination with chemotherapy as a cancer treatment is still being examined and various trials into this subject matter are ongoing. This systematic review summarizes the clinical trials of IV ascorbate to date which were primarily composed of single-arm trials examining dose-limiting toxicities, progression-free survival, and overall survival.

1.1. Clinical Pharmacokinetics of Vitamin C

Clinical data show that intravenous and oral administration of ascorbate yield differing plasma concentrations. When ascorbate is given orally, fasting plasma concentrations are maintained at <100 μM [13] but when oral doses exceed 200 mg, the percentage of the absorbed dose decreases, with a decrease in ascorbate bioavailability, and renal excretion increases [13,19]. In contrast, intravenous administration bypasses the intestinal absorption system. This allows plasma concentrations to be elevated to pharmacologic concentrations (mmol/L [mM] values) that are unachievable via oral administration [20]. In healthy humans, plasma vitamin C concentrations were significantly higher following IV administration compared to oral dosing, and the difference in plasma concentration increased according to the dose delivered. It was found that the mean peak values from IV administration were 6.6-fold higher than the mean peak values from oral administration at a dose of 1.25 g vitamin C [16]. IV ascorbate can be administered by either bolus or continuous infusion. Bolus infusion can be considered as dosing based on pharmacokinetics that occurs over a defined period of time, usually 1.5–2 h [17,21,22]. Continuous infusion is usually considered over periods of time >12 h. Bolus administration of ascorbate has been used more commonly than continuous administration. With a dose of 1 g/kg, bolus administration produces peak plasma ascorbate concentrations of approximately 25 mM, with concentrations maintained above 10 mM for approximately 4 h and return to baseline (<0.1 mM) after approximately 12 h [18]. Following IV administration of pharmacologic ascorbate doses, the plasma half-life is as rapid as 0.5–1 h. With 10 g administered continuously over 24 h, steady-state plasma concentrations can be estimated to be approximately 1–2 mM [19,23]. When oral ascorbate intake stops, the plasma half-life is approximately 8–20 days, due to the action of renal transporters reabsorbing filtered ascorbate [9,18,24,25].

Additionally, some but not all preclinical data indicate that ascorbate can accumulate in solid tumors at higher concentrations than surrounding normal tissue [26–28]. This suggests that cancerous cells are especially affected by vitamin C, which favors the clinical potential of high-dose intravenous vitamin C as a cancer therapeutic [20].

1.2. Possible Mechanisms of Anti-Tumor Effects of Vitamin C

Several major mechanisms have been proposed to explain why only pharmacologic ascorbate concentrations have cytotoxic effects on some but not all cancer cells. Two mechanisms include increased pro-oxidant damage that is irreparable by tumor cells, and oxidation of ascorbate into dehydroascorbic acid (DHA), which is an unstable metabolite and can be cytotoxic [20]. Most data indicate that the first pathway predominates, specifically by generation of extracellular hydrogen peroxide (H_2O_2) by pharmacologic ascorbate and a trace transition metal, usually iron [29,30]. Hydrogen peroxide is cell permeant, and, in the presence of pharmacologic ascorbate, H_2O_2 reactive oxygen species (ROS) are formed extracellularly and/or intracellularly [31]. These ROS have multiple downstream targets, including but not limited to DNA damage, mitochondrial damage, and stimulation of apoptotic pathways [29,32,33].

To learn experimentally whether extracellular H_2O_2 is essential, the enzyme catalase is added. At concentrations used by nearly all laboratories, catalase is a non-permeant protein that dismutates H_2O_2 to water and oxygen. The great majority of in vitro work shows that cell death is blunted or eliminated by catalase addition, pointing to the key role of H_2O_2. The second pathway involves dehydroascorbic acid (DHA), the reversible oxidized form of ascorbate. This pathway is based on findings that tumor cells transport DHA and then internally reduce it to ascorbate. In specifically engineered cells, this reduction triggers scavenging of glutathione (GSH), induces oxidative stress, inactivates glyceraldehyde 3-phosphate dehydrogenase (GAPDH), inhibits glycolytic flux, and leads to an energy crisis that triggers cell death [34,35]. DHA findings are attractive, but have several limitations, including that extracellular H_2O_2 may still be the initial driver of ascorbate oxidation to DHA, and that DHA does not cause cell death in a variety of unmodified cancer cells that do respond to ascorbate [29,30,36,37].

Two additional mechanisms of ascorbate action in cancer are based on ascorbate's activity as a cofactor for Fe (II) 2-oxoglutarate dioxygenase enzymes. As a co-factor, ascorbate modulates DNA demethylation and epigenetic marks through interaction with the ten eleven translocation (TET) enzyme family [38,39]. Ascorbate binds to the catalytic domain facilitating TET-mediated DNA demethylation [38,40]. This reverses the hypermethylation triggered in oncogenic states and subsequently activates tumor suppressor genes [40,41]. Reactivation of tumor suppressor genes allows for anti-tumor mechanisms to become active and increases chemosensitivity. Ascorbate action on TET may have promise in preventing tumor development especially in myelodysplastic syndrome [3,42]. Similarly, ascorbate acts as a co-factor for hypoxia-inducible transcription factors (HIFs) prolyl-4-hydroxylase domain (PHD) enzymes. Prolyl-4-hydroxylation is necessary for targeting of HIFs for proteolytic degradation [43–45]. In solid tumors, HIF-1 helps tumor cells shift from aerobic metabolism to anaerobic metabolism increasing flux through glycolysis to maintain energy production [43]. This activity in tumor cells creates a state that is dependent on glycolytic metabolites. It is possible that the DHA mechanism discussed above works in tandem with the HIF mechanism to cause global disruption of metabolic functioning in the tumor cell triggering cell death. For both TET-mediated and HIF-mediated mechanisms, ascorbate action at physiologically relevant concentrations may prevent cancer development. For cancer treatment, only pharmacologic ascorbate was found to be effective [30].

For the majority of cancer cells in vitro, ascorbate concentrations less than 5 mM are sufficient to induce a 50% decrease in cell survival. In contrast, many non-cancerous cells are capable of tolerating ascorbate concentrations of 20 mM, indicating less sensitivity [36]. Note that in vitro there is some heterogeneity in response to ascorbate in tumor and non-tumor cells alike. Perhaps 10–15% of cancer cells are insensitive to 20 mM ascorbate. Moreover, the death of cancer cells is thought to be selectively induced by extracellular ascorbate, and not intracellular ascorbate [17,36,46,47].

1.3. Synergy with Chemotherapy

Translational synergy of pharmacologic ascorbate with chemotherapy was first demonstrated using cell and mouse pancreatic cancer models [48]. Ascorbate was synergistic with gemcitabine both in vitro and in vivo, without apparent harm. The synergy of ascorbate with conventional chemotherapy is the subject of many clinical studies (Tables 1 and 2). Further, ascorbate was permissive for dose reductions of gemcitabine in these pre-clinical studies. These findings have clinical promise, but to date only individual cases have been reported, without data for failure rates [49]. Ascorbate synergy with conventional chemotherapy was also rigorously investigated in ovarian cancer models. The combination of ascorbic acid and conventional chemotherapeutic agents synergistically inhibited ovarian cancer cell lines and xenografts in mice [50]. Ma et al. exposed ovarian cancer cell lines (OVCAR5, OVCAR8, and SHIN3) to ascorbate and carboplatin in varying molar ratios, using HIO-80 cells, a nontumorigenic ovarian cell line, as a control. The results of this preclinical study demonstrated that the combination of ascorbate and carboplatin induced greater cell death in all cancer cell lines compared to either drug individually [50]. The HIO-80 ovarian epithelial cell line was shown to be equally sensitive to carboplatin alone, and the ascorbate-carboplatin combination. The SHIN3 cell line was implanted into athymic mice to further test the synergistic effect. Ascorbate and carboplatin were shown to be more effective at reducing tumor burden compared to either ascorbate or carboplatin alone. Clinically, multiple trials have demonstrated the safety of ascorbic acid when combined with chemotherapy in the treatment of several cancers including multiple myeloma, ovarian and pancreatic cancer [21,50–52].

2. Materials and Methods

This review's protocol was developed by the authors and was designed to summarize the results of clinical trials in which cancer patients are treated with intravenous vitamin C, either as a single agent or in combination with standard therapies. The population of interest for this review included patients with a current diagnosis of cancer of any type and stage. The intervention of interest was treatment with intravenous ascorbate alone or in combination with standard cancer therapies. Uncontrolled studies or controlled studies that included comparisons against no treatment, placebo, or other standard of care therapies were of interest. Outcomes of interest included Common Terminology Criteria for Adverse Events (CTCAE) adverse events or other measured toxicities, quality of life, progression free survival and overall survival. Randomized controlled trials were of primary interest, but all study designs were included in the initial search.

An electronic literature search was conducted in the PubMed, MEDLINE, and Cochrane databases. PubMed served as an interface for searching MEDLINE (Figure 1). The exact search term combination used in the PubMed search was: "Ascorbate OR Vitamin C AND Cancer NOT Bowel Preparation AND Clinical Trial". The exact search term combination used in the Cochrane database search was: "cancer" "vitamin c" "clinical trial". Following retrieval of the studies, three authors screened the studies (GN, RP, and CJP), eliminated duplicates, and removed all studies that were not clinical trials or not relevant to the subject matter. These authors then screened the remaining studies a second time, removing trials that examined oral ascorbate, and trials that were terminated prematurely and/or had no results. After the second screening, the remaining studies were summarized using a narrative approach.

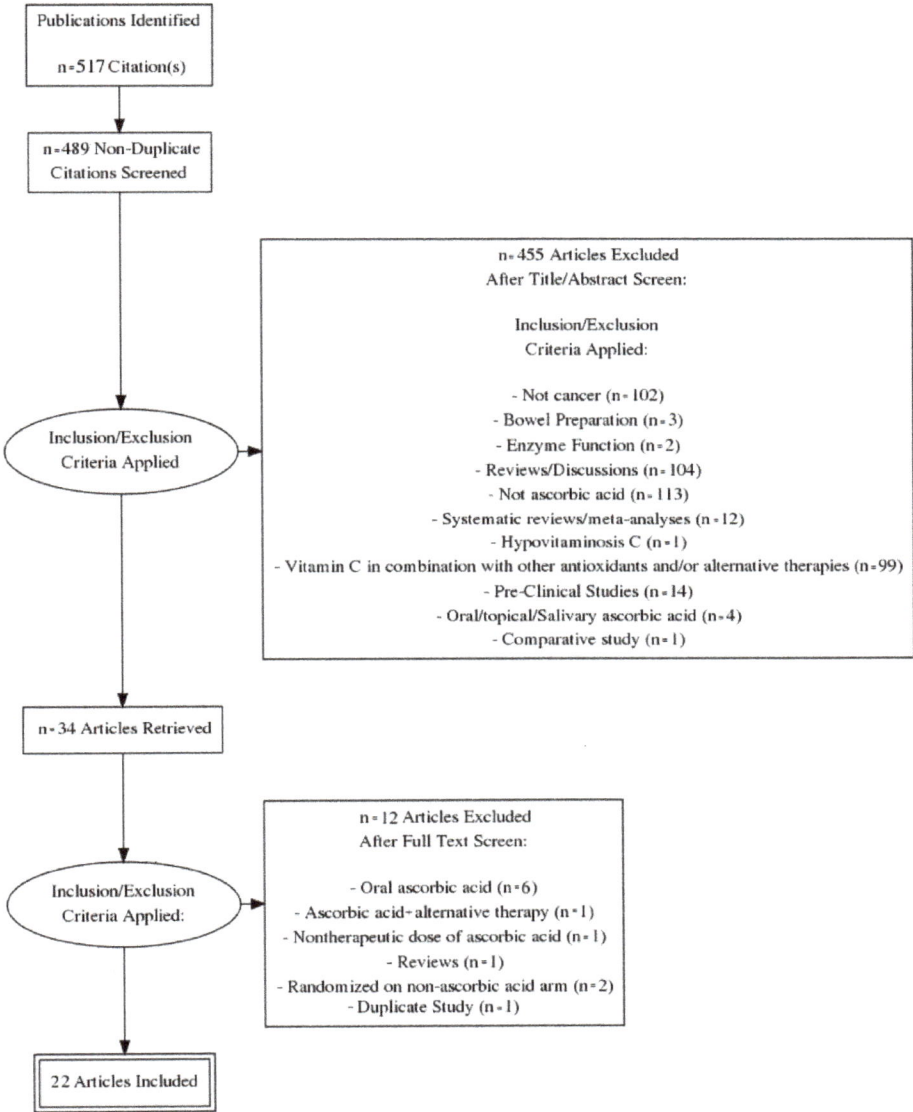

Figure 1. Prisma Flow Diagram.

Study Characteristics

A total of 22 articles (containing 23 trials) that included 401 patients evaluated IV ascorbate (Tables 1–3). Of these trials, eleven trials evaluated arsenic trioxide in combination with intravenous ascorbic acid clinical trials, nine evaluated intravenous ascorbic acid in combination with non-redox cycling agents, and three trials evaluated intravenous ascorbic acid alone. The median sample size of these studies was 17 (range, 3–65) and the IV dose of ascorbate ranged from 1 g daily to 1.5 g/kg thrice weekly.

Table 1. Low dose IV ascorbate + arsenic trioxide trials—Phase I and II trials.

Reference	n	Patient Diagnosis	Trial Design	IV AA Treatment Type and Frequency	Concurrent Treatment Dose	Toxicity	Reported Outcomes/Conclusions
[52]	22	Refractory multiple myeloma	Single Arm	1 g on days 1, 4, 8, and 11 of a 21-day cycle for a maximum of 8 cycles	Bortezomib and Arsenic Trioxide	One occurrence of grade 4 thrombocytopenia was observed in a patient receiving high-dose bortezomib	Objective responses were observed in 27% of patients (2 partial and 4 minor). Median progression-free survival was 5 months and overall survival had not been reached.
[53]	65	Relapsed or refractory multiple myeloma	Single Arm	1 g on days 1–4 of week 1 and twice weekly during weeks 2–5 of a 6 week cycle.	Melphalan and Arsenic Trioxide	Grade 3/4 hematological (3%) or cardiac adverse events occurred infrequently, but grade 3/4 adverse events fever/chills (15%), pain (8%), and fatigue (6%) were reported.	Objective responses occurred in 48% of patients, including complete, partial, and minor responses. Median progression-free survival and overall survival were 7 and 19 months respectively.
[54]	20	Multiple myeloma, relapsed and refractory	Single Arm	1000 mg for 5 consecutive days during week 1, followed by twice weekly during weeks 2–12	Dexamethasone and Arsenic Trioxide	Grade 3 events in 45% and grade 4 events in 5%	30% complete and partial response. Overall median survival was 962 days. 10 patients developed grade 3/4 toxicity to combination treatment.
[55]	17	Lymphoid malignancies, relapsed and refractory.	Single Arm	1000 mg for 5 days during week 1 followed by twice weekly during weeks 2–6	Arsenic Trioxide	1 cardiac death, multiple grade 3 and 4 events	Overall median survival was 7.6 months 6% complete and partial response. Study closed at first interim analysis.
[56]	11	Advanced melanoma	Single Arm	1000 mg for 5 days during week 0, and then twice weekly for an 8 week cycle.	Temozolomide and Arsenic Trioxide	Multiple grade 1 and 2 events.	No responses seen in the first 10 evaluable patients leading to early closure of study.
[57]	5	Refractory metastatic colorectal carcinoma	Single Arm	1000 mg/day for 5 days a week for 5 weeks	Arsenic Trioxide	Grade 3 nausea, vomiting, diarrhea, thrombocytopenia, and anemia	No complete or partial remission observed. CT scans showed stable or progressive disease.
[58]	20	Multiple myeloma, relapsed and refractory	Single Arm	1 mg (one dose during the first week, twice weekly during weeks 2–4)	Dexamethasone and Arsenic Trioxide	Multiple grade 3 and 4 events	Clinical response was observed in 40% of patients (including partial and minor). Median progression free survival = 4 months and median overall survival = 11 months. Authors state that it was difficult to assess activity of each individual agent.
[59]	11	Non-acute promyelocytic leukemia; acute myeloid leukemia (non-APL AML)	Single Arm	1 g/day for 5 days a week for 5 weeks	Arsenic Trioxide	Few grade 3 or 4 adverse effects and the most common grade 3 toxicity was infection though possibly related to the leukemia	One patient achieved a complete response; another achieved a complete remission with incomplete hematologic recovery. Authors concluded that arsenic trioxide + ascorbic acid had limited clinical meaning in non-APL AML patients.
[60]	6	Relapsed or refractory myeloma	Single Arm	1000 mg/day for 25 days over 35 days total.	Arsenic Trioxide	One episode of grade 3 hematologic toxicity (leukopenia) was observed.	Two patients had partial responses; four had stable disease.

Table 1. *Cont.*

Reference	n	Patient Diagnosis	Trial Design	IV AA Treatment Type and Frequency	Concurrent Treatment Dose	Toxicity	Reported Outcomes/Conclusions
[61]	10	Relapsed/refractory multiple myeloma	Single Arm	1 g daily for 3 days of week 1, then twice weekly for a 3-week cycle.	Arsenic Trioxide and Bortezomib	No dose limiting adverse effects.	40% response rate with one patient achieving a durable partial response.
[62]	13	Myelodysplastic Syndrome and Acute Myeloid Leukemia (concurrent diagnoses)	Single Arm	1 g for 5 days during week following each dose of IV Arsenic Trioxide and then once weekly thereafter	Decitabine and Arsenic Trioxide	Grade 3 and 4 events; two patient deaths occurred not related to treatment	One morphologic complete remission was observed. Five patients had stable disease after recovery. 0.2 mg/kg identified as maximum tolerated dose of arsenic in combination with Decitabine and Ascorbic Acid.

Note: This table illustrates the eleven clinical trials that evaluated intravenous ascorbate in combination with arsenic trioxide.

Table 2. High dose IV ascorbate + standard therapies—Phase I and II Trials.

Reference	n	Patient Diagnosis	Trial Design	IV AA Treatment Type and Frequency	Concurrent Treatment Dose	Toxicity	Reported Outcomes/Conclusions
[63]	17	Advanced tumors	Single Arm	Five cohorts treated with 30, 50, 70, 90, and 110 g/m² for 4 consecutive days for 4 weeks.	Multivitamin and Eicosapentaenoic acid	Grade 3 and grade 4 hyponatremia, hyperkalemia	3 patients had stable disease, 13 had progressive disease. Recommended dose is 70–80 g/m². This translates to approximately 125 g because the average patient has a body surface area of 1.6–1.9 m².
[64]	3	Relapsed lymphoma	Single Arm	75 g twice weekly	Rituximab, cyclophosphamide, cytarabine, etoposide, dexamethasone	Grade 3 neutropenia, anemia, thrombocytopenia	The authors concluded that 75 g was a safe dose.
[51]	11	Advanced pancreatic adenocarcinoma	Single Arm	15–125 g twice weekly	Gemcitabine	No dose limiting adverse effects	Mean plasma ascorbate levels were significantly higher than baseline. Mean survival time of subjects completing 8 weeks of therapy was 13 ± 2 months.
[21]	14	Pancreatic adenocarcinoma, stage IV	Single Arm	50, 75, and 100 g per infusion (3 cohorts) thrice weekly for 8 weeks	Gemcitabine and Erlotinib	Multiple toxicities, all grades, thought to not be related to AA; grade 4 adverse event included two patients with pulmonary embolism	50% of patients had stable disease. Survival analysis excluded 5 patients who progressed quickly (3 died). Overall mean survival was 182 days.

Table 2. *Cont.*

Reference	n	Patient Diagnosis	Trial Design	IV AA Treatment Type and Frequency	Concurrent Treatment Dose	Toxicity	Reported Outcomes/Conclusions
[58]	25	Stage 3/4 ovarian cancer	Randomized	75 or 100 g twice weekly for 12 months (target plasma concentration 20–23 mM)	Carboplatin and paclitaxel	Ascorbate did not increase grade 3/4; grade 1 and 2 toxicities were substantially decreased	8.75 month increase in PFS in AA-treated arm. Trend to improved OS in AA group; no numerical data reported.
[22]	16	Various cancer types (lung, rectum, colon, bladder, ovary, cervix, tonsil, breast, biliary tract)	Single Arm	1.5 g/kg body weight infused three times (at least one day apart) on week days during weeks when chemotherapy was administered (but not on the same day as intravenous chemotherapy) and any two days at least one day apart during weeks when no chemotherapy was given.	Standard care chemotherapy.	Increased thirst and increased urinary flow; these adverse symptoms did not appear to be caused by the ascorbate molecule	Patients experienced unexpected transient stable disease, increased energy, and functional improvement.
[70] Phase I study	13	Glioblastoma	Single Arm	Radiation phase: radiation (61.2 Gy in 34 fractions), temozolomide (75 mg/m² daily for a maximum of 49 days), ascorbate (doses ranging from 15–125 g, 3 times per week for 7 weeks) Adjuvant phase: 6 cycles of 28 days; treatment with temozolomide (1 dose-escalation to 200 mg/m² if no toxicity in cycle 1), ascorbate (2 times per week, dose-escalation until 20 mM plasma concentration, around ~85 g infusion).	Ascorbate with radiation and temozolomide	Radiation phase toxicity: Grade 2 and 3 fatigue and nausea; grade 2 infection; grade 3 vomiting Adjuvant phase toxicity: grade 2 fatigue and nausea; grade 1 vomiting; grade 3 leukopenia; and grade 3 neutropenia.	Progression-free survival 13.3 months; average overall survival 21.5 months.
[30] Phase II study	14	Advanced stage non-small cell lung cancer	Single Arm	1 cycle is 21 days; IV carboplatin (AUC 6, 4 cycles), IV paclitaxel (200 mg/m², 4 cycles), IV pharmacological ascorbate (two 75 g infusions per week, up to 4 cycles)	Carboplatin, paclitaxel, and ascorbate	No grade 3 or 4 toxicities related to ascorbate	Imaging confirmed partial responses to therapy (n = 4); stable disease (n = 9), disease progression (n = 1)
[65]	14	Locally advanced or metastatic prostate cancer	Single Arm	Phase I: Escalating dose of IVC from 25 g to 100 g and gemcitabine alone at 1000 mg/m² (week 3) with a few patients receiving reduced doses and gemcitabine with IVC (week 4) Phase IIa: no gemcitabine for 1 week and then continuous treatment of gemcitabine until disease progression or unacceptable toxicity and IVC 3 times per week	IVC and gemcitabine	Low toxicity; Increased thirst and nausea were caused by IVC	Patients experienced a mix of stable disease, partial response and disease progression.

Note: This table illustrates the nine clinical trials that evaluated intravenous ascorbate in combination with non-redox cycling chemotherapy agents.

Table 3. High dose IV ascorbate only—Phase I and II trials.

Reference	n	Cancer Type	Trial Design	IV AA Treatment Type and Frequency	Toxicity	Reported Outcomes/Conclusions
Phase I						
[18]	24	Advanced cancer or hematologic malignancy	Single Arm	1.5 g/kg body weight three times weekly	No dose limiting adverse effects.	Two patients had unexpectedly stable disease.
Phase II						
[66]	23	Castration-resistant prostate cancer	Single Arm	5 g during weekly week 1, 30 g weekly during week 2, and 60 g weekly during weeks 3–12	Multiple grade 3 events including hypertension and anemia; two patients experienced pulmonary embolism.	Adverse events were thought to be more likely related to disease progression than ascorbic acid.
[23]	11	Late stage terminal cancer patients	Single Arm	150–710 mg/kg/day for up to eight weeks	Two Grade 3 adverse events: one patient with a history of renal calculi developed a kidney stone after thirteen days of treatment and another patient experienced hypokalemia after six weeks of treatment.	One patient had stable disease and continued the treatment for forty-eight weeks Intravenous vitamin C was deemed relatively safe so long as the patient does not have a history of kidney stone formation.

Note: This table illustrates the three IV ascorbate-only trials evaluated in this review. These trials evaluated IV ascorbate as a single intervention.

3. Results

3.1. Trials Evaluating Low-Dose Intravenous Ascorbic Acid in Combination with Arsenic Trioxide

Out of the 23 clinical trials included in this paper, 11 trials with a total of 200 patients used low dose intravenous ascorbic acid in combination with arsenic trioxide (As_2O_3) (Table 1). In the study design of trials such as these, ascorbic acid does not act as an anti-cancer therapy in its own right. The dose used in such studies (one gram per day) is not considered a pharmacologic or effective dose. The justification for selecting this dose is that when given orally it saturates plasma and ascorbate tissue ascorbate concentrations. However, in most studies, ascorbate was administered intravenously, for unclear reasons [29]. Ascorbate was added as a redox cycling compound to facilitate the anti-cancer activity of As_2O_3. Thus, for these trials, the main source of anti-cancer activity is As_2O_3 and all of the effects produced in these trials should be attributed to As_2O_3 [29].

Berenson et al. reported two phase II trials [52,53] in patients with refractory/multiple myeloma that included low dose IV ascorbic acid. In the 2006 study, patients (n = 65) received IV ascorbic acid (1 g on Days 1–4) and As_2O_3 and melphalan. A response rate of 48% was observed with a progression-free survival of seven months and overall survival of 19 months [53]. In the 2007 study, patients (n = 22) were treated with IV ascorbate (1 g on Days 1, 4, 8, and 11 of a 21-day cycle for a maximum of eight cycles) in combination with Bortezomib and As_2O_3 [52]. A response rate of 27% was observed and median progression-free survival was five months. Both studies reported grade 3/4 adverse events. Because of the trial design and types of toxicities, adverse events were likely related to chemotherapy.

Two As_2O_3 trials examined the benefits of IV ascorbate in combination with chemotherapy used response rate as the primary outcome [54,55]. Abou-Jawade et al. reported a single arm study of IV ascorbate (1000 mg daily for five days and then twice weekly for nine weeks) in combination with Dexamethasone and As_2O_3 for patients of relapsed and refractory myeloma (n = 20). The authors reported an overall response rate of 30%, which included both partial and complete response. Ten patients developed grade 3 or 4 toxicity to this treatment combination, although toxicity due to ascorbate was not defined. Chang et al. reported a similar phase II trial in which patients with lymphoid malignancies (n = 17) were treated with IV ascorbate (1000 mg daily for five days then twice weekly) alongside As_2O_3. An overall response rate of 6% was reported and severe toxicities (multiple grade 3, 4, and 5 events) were observed. The trial was closed after the first interim analysis due to lack of activity. Similarly, in a phase II trial by Bael et al. in patients with advanced melanoma (n = 11) being treated with IV ascorbate (1000 mg for five days for one week, then twice weekly for an additional eight weeks) in combination with Temozolomide and As_2O_3, no responses were seen in the first 10 evaluable patients leading to early closure of the study [56].

Three trials examined the benefit of IV ascorbate (1000 mg/day for five days) in combination with As_2O_3 only [57–59]. Subbarayan et al. reported a study in which patients of refractory metastatic colorectal carcinoma (n = 5) were treated with this combination, and multiple grade 3 events were reported (nausea, vomiting, diarrhea, thrombocytopenia, and anemia), although no complete or partial remission was observed [57]. Wu et al. reported a similar trial with patients of relapsed/refractory multiple myeloma (n = 20), but this study was reported in a letter format only [58]. A median survival time of 11 months was observed. In the 2014 study by Aldoss et al., however, intravenous AA was evaluated in combination with As_2O_3, which is highly effective in acute promyelocytic leukemia (APL), but, despite its multiple mechanisms of action, it has no activity in acute myeloid leukemia (AML) that excludes APL (non-APL AML). The patient population (n = 11) in this study were all diagnosed with non-APL AML and were administered intravenous As_2O_3 (0.25 mg/kg/day over 1–4 h) with intravenous AA (1 g/day over 30 min after As_2O_3) for five days a week for five weeks (25 doses). Among 10 evaluable patients, one achieved a complete response, one achieved a partial remission with incomplete hematologic recovery, and four patients had disappearance of blasts from peripheral blood and bone marrow. The observed As_2O_3 toxicity was mild; very few grade 3 or 4 adverse effects and the most common grade 3 toxicity was infection, although possibly related to the leukemia. The authors concluded that combination of As_2O_3 and AA had limited clinical meaningful anti-leukemia activity in patients with non-APL AML [59].

Bahlis et al. reported a study using As_2O_3 in combination with IV ascorbate to ascertain dosing of As_2O_3. Patients of refractory myeloma (n = 6) were treated with IV ascorbate dose of 1000 mg/day for 25 days over 35 days total, and 0.25 mg/kg per day of As_2O_3 was defined as an appropriate dose [60]. A partial response rate of 36% was observed and no toxicities above grade 2 were reported. It is unclear if these toxicities were due to the addition of ascorbic acid, increased As_2O_3, the schedule, or duration of treatment. Held et al. reported a similar phase I trial that also aimed to estimate the maximum tolerated dose of As_2O_3 and bortezomib that can be used in combination with IV ascorbate (1 g daily for three days of Week 1, then twice weekly for a three-week cycle) in patients with relapsed/refractory multiple myeloma (n = 10) [61]. No dose-limiting toxicities were reported and a 40% response rate was reported. Welch et al. reported a trial with patients of myelodysplastic syndrome and acute myeloid leukemia (n = 13) being treated with Decitabine and As_2O_3 and IV ascorbate (1000 mg for five days during Week 1 following each dose of IV As_2O_3 and then once weekly thereafter) [62]. Five patients had stable disease after recovery and multiple grade 3 and 4 events were reported; the authors stated that these adverse events were expected given the patient population and type of chemotherapy but did not clarify if the addition of ascorbate was a contributing factor.

3.2. Trials Evaluating High-Dose Intravenous Ascorbic Acid with Standard Chemo- and Radiotherapy Agents

Stephenson et al. reported a phase I trial with patients with advanced malignancies (n = 17) being treated with IV ascorbate 70–80 g/m^2 (this translates to approximately 125 g because the average patient has body surface of 1.6–1.9 m^2) [67] 3–4 times a week to obtain optimal peak plasma concentrations in combination with multivitamin and eicosapentaenoic acid treatment [63]. Only two patients completed the entire four-week study period, and stable disease rate and progressive disease rate of 19% and 81% were reported, respectively. Grade 3 and 4 metabolic toxicities (hypernatremia and hypokalemia) related to ascorbate was observed. Kawada et al. reported a similar study in patients with relapsed lymphoma (n = 3) that were treated with rituximab, cyclophosphamide, cytarabine, etoposide, and dexamethasone alongside IV ascorbate (75 g twice weekly) [64]. Grade 3 neutropenia, anemia, and thrombocytopenia were observed, but no obvious side effects due to ascorbic acid were observed, leading the authors to conclude that 75 g of IV ascorbate is a safe dose. It is likely that hypernatremia and hyperkalemia, reported by Stephenson et al., was secondary to the approximately two-fold higher ascorbate dose that patients received in comparison to other trials.

Two phase I trials examined the benefits of IV ascorbate in combination with gemcitabine in patients with advanced pancreas adenocarcinoma [21,51]. Both Welsh et al. (n = 13) and Monti et al. (n = 14) reported toxicity in patients related to gemcitabine and not secondary to ascorbate [21,51]. Response rates and survival duration in both studies were reported only for patients who did not progress within the first month of treatment and are thus not representative of standard clinical reporting. Monti et al. reported that seven of nine patients had stable disease with a mean overall survival of 155 ± 182 days and Welsh et al. reported a 13 ± 2-month mean survival in the nine patients that were analyzed.

Ma et al. reported a trial of patients with stage 3 and 4 ovarian cancer (n = 25) receiving carboplatin and paclitaxel chemotherapy [50]. Patients were randomized to either IV ascorbate (75 g or 100 g twice weekly for 12 months) with chemotherapy (n = 13) or chemotherapy alone (n = 12). The trial was not blinded and the primary outcome was toxicity. The ascorbate group was observed to have fewer grade 1/2 adverse events per encounter as compared to the group that received only chemotherapy. A trend toward improvement in median overall-survival was reported, although no numerical data were reported. Median time for disease progression/relapse was reported as 25.5 months in the ascorbate arm and 16.75 in the chemotherapy arm. This trial also demonstrated key information related to the safety profile of ascorbate as patients were treated for more than a year with minimal adverse effects.

Hoffer et al. (2015) reported a study with patients of various cancer types (*n* = 16) treated with IV ascorbate. Patients were administered ascorbate at a dose of 1.5 g/kg three times on weekdays during weeks when chemotherapy was administered, and at least one day apart during weeks when no chemotherapy was given). This was given in combination with standard care chemotherapy, which was not defined [22]. Adverse effects included increased thirst and urinary flow. Transient stable disease, increased energy, and functional improvement were observed in patients.

Schoenfeld et al. (2017) reported a phase I study with glioblastoma (GBM) patients (*n* = 13) receiving pharmacological ascorbate with radiation and temozolomide [30]. The study had two phases: the radiation phase (which started on Day 1 of the radiation phase and ended on Cycle 1, Day 1 of the adjuvant period) and the adjuvant phase (which began on Cycle 1, Day 1 until Cycle 6, Day 28) [30]. The participants in the radiation phase received radiation (61.2 Gy in 34 fractions), temozolomide (75 mg/m^2 daily for a maximum of 49 days) and ascorbate (dose cohorts ranging from 15–125 g, three times per week for seven weeks) [30]. In the adjuvant phase, participants received temozolomide (Days 1–5 of a 28-day cycle and one dose-escalation to 200 mg/m^2 took place if Cycle 1 was tolerable) and ascorbate (infusions took place two times per week and dose was increased over two infusions until a plasma concentration of 20 mM was reached, which was achieved with an 87.5 g infusion) for about 28 weeks [30]. Adverse effects in the radiation phase included grade 2 and 3 fatigue and nausea, grade 2 infection, and grade 3 vomiting. In the adjuvant phase, patients experienced grade 2 fatigue and nausea, grade 1 vomiting, grade 3 leukopenia, and neutropenia. At the time of publishing in 2017, the average PFS with Schoenfeld et al.'s therapy was 13.3 months as compared to PFS of seven months in Stupp et al. (2005) [61] which treated GBM patients with similar characteristics with concurrent radiation and temozolomide or radiation only. Average overall survival was 21.5 months as compared to 14 months in Stupp et al., 2005 [30].

Schoenfeld et al. (2017) also reported a phase II study with advanced stage non-small-cell-lung carcinoma (NSCLC) patients (*n* = 14) treated with carboplatin, paclitaxel and pharmacological ascorbate [30]. Participants were administered IV carboplatin (AUC 6, four cycles), IV paclitaxel (200 mg/m^2, four cycles), and IV pharmacological ascorbate (75 g per infusion, two infusions per week, up to four cycles); one cycle was 28 days [30]. No grade 3 or 4 toxicities related to ascorbate were noted. Imaging-confirmed partial responses to therapy in patients who completed the trial (*n* = 4), stable disease (*n* = 9), and new lesion development (*n* = 1) indicating disease progression despite the patient having a stable target lesion [30].

Polireddy et al. (2017) reported a Phase I/IIa trial with locally advanced or metastatic prostate cancer patients (*n* = 14) who were not eligible for surgical resection with high-dose IVC and gemcitabine chemotherapy [62]. Phase I initially enrolled 14 patients but only 12 patients completed a pharmacokinetic evaluation of IVC and gemcitabine alone. IVC dose escalated from 25 g to 100 g and gemcitabine dose at 1000 mg/m^2, with a few patients receiving reduced doses as determined by the treating oncologist give from Week 1 to Week 4 and subsequently in combination during Week 4. In Phase IIa, the 12 patients were given IVC three-times weekly at doses determined by the treating oncologist and gemcitabine following a rest week after two consecutive weeks of a determined dose and then treatment until tumor progression or patient withdrawal. Overall survival was 15.1 months with 5 of 12 patients not surviving over one year, 6 of 12 patients surviving over one year, and 1 of 12 surviving more than two decades after diagnosis. Over the course of treatment, one patient with Stage III pancreatic ductal carcinoma experienced tumor shrinkage/stabilization and tumor margins becoming more distinct, making the patient eligible for surgery. Grade 1 nausea and thirst related to IVC were the only adverse events noted. The study showed IVC has low toxicity and does not alter gemcitabine pharmacokinetics significantly.

3.3. IV Ascorbate Only Trials

Hoffer et al. (2008) reported a phase I trial in patients with advanced cancer or hematologic malignancy (*n* = 24) treated with up to 1.5 g/kg body weight of IV ascorbate three times weekly.

No dose limiting adverse effects were reported and two patients had unexpectedly stable disease [18]. Nielsen et al. (2017) reported a similar study in patients with castration-resistant prostate cancer (*n* = 23) treated with IV ascorbate 5 g once during Week 1, 30 g weekly during Week 2, and 60 g once weekly during Weeks 3–12 [34]. Multiple grade 3 events were reported including hypertension and anemia; two patients experienced a pulmonary embolism; however, the authors stated that treatment-induced toxicity was limited and the two episodes of pulmonary embolism can likely be attributed to the fact that cancer is known to increase the risk of thromboembolic events. However, without a placebo-controlled trial, attribution to disease or ascorbate cannot be definitely proven. Both studies reported no anticancer response or disease remission.

Lastly, Riordan et al. reported a trial in which late stage terminal cancer patients (*n* = 11) were given continuous infusions of 150–710 mg/kg/day for up to eight weeks. Intravenous infusions increased plasma ascorbate concentrations to a mean of 1.1 mM. Two Grade 3 adverse events to the agent were reported: one patient with a history of renal calculi developed a kidney stone after thirteen days of treatment and another patient experienced hypokalemia after six weeks of treatment; the authors state that these adverse events could possibly be related to ascorbic acid, but it remains unclear. One patient had stable disease and continued the treatment for forty-eight weeks. The authors concluded that intravenous vitamin C administered continuously is relatively safe so long as the patient does not have a history of kidney stone formation [23].

3.4. Potential of Benefit and Current Limitations

Clinical trials that have examined the use of IV ascorbate in cancer patient populations have yielded results that suggest its potential to produce various beneficial effects. In one trial, IV ascorbate was used in elderly patients with advanced cancer who had failed all other therapies. Two of the patients had unexpected stable disease after eight weeks of ascorbic acid treatment [18]. A phase I trial in patients with metastatic stage 4 pancreatic cancer who were treated with gemcitabine and IV ascorbate as primary therapy until tumor progression showed few toxicities associated with the treatment [51]. The nine patients had a tripling of disease free interval compared to literature controls and a doubling of survival compared to retrospective controls. Some patients were treated for longer durations, for instance over a year in the Ma et al. trial [50], and had substantially decreased grade 1 and 2 adverse events when compared to the group not receiving ascorbate.

Similarly, a trial of patients with metastatic stage 4 pancreatic cancer illustrated benefit as eight of nine patients experienced tumor shrinkage after eight weeks of primary therapy (gemcitabine and erlotinib) and pharmacological ascorbate as measure by CT scans [21]. The results of these trials and others discussed in this article suggest that IV ascorbate is useful as a single agent or combined with a primary therapy. It has the benefit of being a non-toxic treatment modality and reducing toxicity of chemotherapeutics when combined with conventional therapies. In one randomized ovarian cancer trial, patients receiving IV ascorbic acid reported lower levels of low-grade gastrointestinal, hepatobiliary, dermatological, immune/infection, pulmonary and renal toxicities commonly associated with carboplatin and paclitaxel treatment [50]. One retrospective cohort study [68] compared breast cancer patients who received IV ascorbic acid to those who did not, at a dose of 7.5 g weekly without blinding. In the first year following surgery, patients who received ascorbate when compared to a control group had significant reductions in nausea (p = 0.022), loss of appetite (p = 0.005), fatigue (p = 0.023), dizziness (p = 0.004) and hemorrhagic diathesis (p = 0.032). Limitations of this study are the absence of blinding, the non-therapeutic dosing and once weekly frequency of ascorbate administration. Even so, the results suggest that IV ascorbate could induce reduction in toxicities, perhaps via mechanisms that are different than those that target cancer cells. To definitively associate IV ascorbate with clinical benefit and/or toxicity, more rigorous randomized-placebo trials must be conducted. Many of the clinical studies conducted to date do not contain a control group, which makes determining efficacy difficult.

In addition to examining efficacy, there is need for a determination of IV AA dosing amount, dosing frequency, and duration of treatment alone or in combination with other therapies. Currently, there is no consensus on these parameters. For dosing of intravenous ascorbate, doses used most frequently are based on one of a few regimens suggested by Riordan and colleagues. The goal was to achieve a plasma concentration of approximately 22 mM, which was effective in a hollow-fiber tumor model [69]. This dosing amount translates as approximately 1 g/kg. For dosing frequency, a regimen of 2–3 times weekly was empiric, with patient ability and/or willingness to receive treatment being limiting factors. Considering dosing amount and frequency together, therapy less than twice weekly, with dosing less than 1 g/kg, appears to be therapeutically ineffective [66], while dosing at 1 g/kg at least twice weekly has promise [21,51]. For duration needed to assess responsiveness, clinically detectable ascorbate action is relatively slow compared to many other cancer therapies. In most reports, a minimum 2–3-month time frame was needed to assess response [17,21,51,69,70]. Due to unknowns about concomitant administration with standard chemotherapies, ascorbate most often has been administered alone, without other chemotherapy on the same day. When ascorbate was administered on the same day as chemotherapy in a series of cases, the clinical response was seemingly enhanced by ascorbate [66,70]. Unfortunately, in this case series, only minimal information was provided about adverse events and non-responders. Based on the totality of available evidence and our experience, some recommendations can be made for future studies. These recommendations include that ascorbate dosing should be 1 g/kg, at a minimum frequency of twice weekly, and with a minimum of a two-month and preferably three-month trial period before efficacy is assessed. However, further research into the potential benefit of IV AA is necessary before well-defined clinical recommendations can be made.

4. Future Directions

Thus far, a total of 185 cancer patients have been treated with IV vitamin C within the clinical trials discussed in this review, excluding those treated with low-dose vitamin C (redox coupling mechanism) (Figure 2). Moreover, there are 11 studies in progress that aim to investigate the clinical efficacy of pharmacological IV ascorbate (Table 4) including a total of 405 patients. There are two randomized studies (NCT03175341 and NCT02516670) and two non-randomized studies (NCT01852890 and NCT01752491). Both hematological and solid organ malignancies are being evaluated. Note that, even though all studies using ascorbate are included, ascorbate dosing may be well below that considered pharmacologic dosing.

In the Phase I trials, the safety of high dose ascorbate is being tested in combination with gemcitabine and radiation therapy (NCT01852890), temozolomide, and radiation therapy (NCT01752491), and gemcitabine, cisplatin, and nab-paclitaxel (NCT03410030). These studies aim to further determine the safety and toxicity of high dose ascorbate, in addition to establishing a pharmacokinetic profile, elucidating the biological responsiveness, and determining its efficacy in reducing side effects of chemotherapy.

In the Phase II trials, ascorbate is being tested in combination with gemcitabine and nab-paclitaxel (NCT02905578), temozolomide and radiation therapy (NCT02344355). The randomized Phase II trials, such as *Docetaxel with or without Ascorbic Acid in Treating Patients with Metastatic Prostate Cancer (NCT02516670)*, provide a vehicle for assessing the clinical effectiveness of ascorbate in a high-quality, placebo-controlled setting. These kinds of trials have the potential to further elucidate any synergistic anticancer effects that IV ascorbate might have when used in combination with chemotherapeutic agents. This has the potential to provide patients with additional treatment options.

Table 4. Upcoming and active interventional trials utilizing pharmacological IV ascorbate.

Phase	Trial Title	Trial Design	IV AA Treatment Type and Frequency	Interventions	Status	Enrollment	NCT Identifier
Phase I	Gemcitabine, Ascorbate, Radiation Therapy for Pancreatic Cancer	Single Arm	50 g–100 g during radiation therapy for 5–6 weeks; escalating dose based on tolerance	Ascorbate Gemcitabine Radiation Therapy	Ongoing, closed to accrual	16	NCT01852890
Phase I	High-Dose Ascorbate in Glioblastoma Multiforme	Single Arm	15 g–87.5 g by IV 3×/week for 12 weeks	Ascorbate Temozolomide Radiation Therapy	Ongoing, closed to accrual	13	NCT01752491
Phase I	High Dose Ascorbic Acid (AA) + Nanoparticle Paclitaxel Protein Bound + Cisplatin + Gemcitabine (AA NABPLAGEM) in Patients Who Have No Prior Therapy for Their Metastatic Pancreatic Cancer	Single Arm	No dosing information provided	Ascorbic Acid Nab-paclitaxel Cisplatin Gemcitabine	Ongoing, actively recruiting participants	36	NCT03410030
Phase II	High-dose Ascorbate for Pancreatic Cancer (PACMAN 2.1)	Single Arm	75 g by IV 3×/week for 4 weeks	Ascorbate Gemcitabine Nab-paclitaxel	Accrual began 28 May 2018	30	NCT02905578
Phase II	High-Dose Ascorbate in Stage IV Non-Small Cell Lung Cancer	Single Arm	75 g by IV 2×/week for up to 12 weeks	Ascorbic Acid Carboplatin Paclitaxel	Ongoing, actively recruiting participants	57	NCT0420314
Phase II	High-Dose Ascorbate in Glioblastoma Multiforme	Single Arm	87.5 g by IV 3×/week during radiation therapy After radiation ascorbate 2×/week	Temozolomide Ascorbic Acid Radiation Therapy	Ongoing, actively recruiting participants	90	NCT02344355
Phase II	Adding Ascorbate to Chemotherapy and Radiation Therapy for NSCLC (XACT-LUNG)	Single Arm	Concurrent phase: 75 g by IV 3×/week for up to 7 weeks Consolidation phase: 75 g by IV 2×/week for two cycles (42 days)	Paclitaxel Carboplatin Ascorbate Radiation Therapy	Ongoing, actively recruiting participants	46	NCT02905591
Phase II	Docetaxel with or Without Ascorbic Acid in Treating Patients with Metastatic Prostate Cancer	Randomized	1 g/kg 3× per week	Docetaxel Ascorbic Acid or Placebo	Ongoing, actively recruiting participants	69	NCT02516670

Table 4. *Cont.*

Phase	Trial Title	Trial Design	IV AA Treatment Type and Frequency	Interventions	Status	Enrollment	NCT Identifier
Phase I/II	Randomized Study to Evaluate the Role of Intravenous Ascorbic Acid Supplementation to Conventional Neoadjuvant Chemotherapy in Women with Breast Cancer	Randomized	1.5 g on day 1 followed by 0.75 g on day 2–4 at each chemotherapy cycle	Ascorbic Acid Placebo	Status Unknown	30	NCT03175341
Phase I/II	Evaluating the Safety and Tolerability of Vitamin C in Patients with Intermediate or High Risk Myelodysplastic Syndrome with TET2 Mutations	Single Arm	50 gm CIVI/24 h x 5 days every 4 week	Ascorbic acid	Accrual begins 26 June 2018	18	NCT03433781

Note: This table illustrates the current and upcoming trials listed on clinicaltrials.gov that are utilizing standard of care chemotherapeutics in combination with pharmacological IV ascorbate.

In addition to these clinical studies, laboratory research has attempted to elucidate the mechanism of action of vitamin C in cancer cells. Yun et al. showed that high dose vitamin C selectively killed KRAS and BRAF mutants in colorectal cancer cells by inducing increased uptake of oxidized vitamin C and targeting the glycolytic pathway [34], although these findings have not been confirmed by others [30,36,48]. Nevertheless, potential remains to identify subtypes of cancer that might benefit from IV pharmacological ascorbate in a clinical setting.

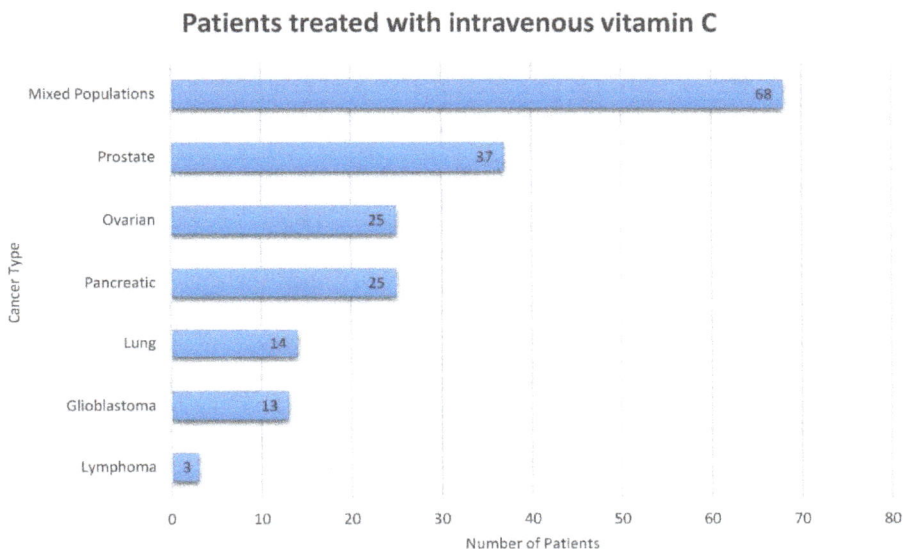

Patients treated with intravenous vitamin C

Figure 2. This figure represents the total number of cancer patients (*n* = 185) that were treated with intravenous ascorbic acid within the clinical trials summarized in this paper. This figure does not include patients who were enrolled in trials that used arsenic trioxide.

5. Conclusions

Evidence supporting the existence of an anticancer effect of intravenous ascorbate is mixed, whether it is given a single agent or in combination with other concurrent standard therapies. In single-arm trials that used IV ascorbate in combination with concurrent standard therapies, it is unclear which agent delivered which effects. Only one randomized clinical trial has been reported, showing a trend toward overall survival, a significant 8.5 week increase in progression-free survival, and decreased adverse events in the IV ascorbate arm in ovarian cancer patients.

Current research indicates that IV ascorbate is well tolerated and has reported some positive results. However, high-quality placebo-controlled trials such as those being offered in prostate (NCT02516670) and breast cancer (NCT03175341) are needed to strengthen the present evidence base to support continuation of IV ascorbate being offered as a treatment by practitioners. Until these trials are completed, patients should be informed of the investigational status of IV ascorbate as a cancer treatment.

Funding: This research was funded by National Institutes of Health: NIH P30 CA006973, K23 CA197526, and NIH DK053212-12.

Acknowledgments: C.J.P. was supported by NIH P30 CA006973, K23 CA197526 and the Marcus Foundation. M.L. was supported by the Intramural Research Program, NIDDK, NIH DK053212-12.

Conflicts of Interest: The authors declare no conflict of interest.

Abbreviations

AA	Ascorbic Acid
AML	Acute Myeloid Leukemia
APL	Acute Promyelocytic Leukemia
As_2O_3	Arsenic Trioxide
CT	Computed Tomography
CTCAE	Common Terminology Criteria for Adverse Events
DHA	Dehydroascorbic acid
DNA	Deoxyribonucleic acid
GBM	Glioblastoma
GADPH	Glyceraldehyde 3-phosphate dehydrogenase
GSH	Glutathione
HIF	Hypoxia inducible factors
H_2O_2	Hydrogen Peroxide
IV	Intravenous
IVC	Intravenous vitamin C
mM	Millimolar
NSCLC	Non-small cell lung cancer
OS	Overall survival
PFS	Progression free survival
PHD	Prolyl-4-hydroxylase domain
ROS	Reactive oxygen species
TET	Ten eleven translocation

References

1. McCormick, W. Cancer: The preconditioning factor in pathogenesis; a new etiologic approach. *Arch. Pediatr.* **1954**, *71*, 313–322. [PubMed]
2. Moertel, C.G.; Fleming, T.R.; Creagan, E.T.; Rubin, J.; O'Connell, M.J.; Ames, M.M. High-dose vitamin C versus placebo in the treatment of patients with advanced cancer who have had no prior chemotherapy. A randomized double-blind comparison. *N. Engl. J. Med.* **1985**, *312*, 137–141. [CrossRef] [PubMed]
3. Cimmino, L.; Dolgalev, I.; Wang, Y.; Yoshimi, A.; Martin, G.H.; Wang, J.; Ng, V.; Xia, B.; Witkowski, M.T.; Mitchell-Flack, M.; et al. Restoration of TET2 Function Blocks Aberrant Self-Renewal and Leukemia Progression. *Cell* **2017**, *170*, 1079–1095. [CrossRef] [PubMed]
4. Cameron, E.; Pauling, L. The orthomolecular treatment of cancer. I. The role of ascorbic acid in host resistance. *Chem. Biol. Interact.* **1974**, *9*, 273–283. [CrossRef]
5. Cameron, E.; Campbell, A. The orthomolecular treatment of cancer. II. Clinical trial of high-dose ascorbic acid supplements in advanced human cancer. *Chem. Biol. Interact.* **1974**, *9*, 285–315. [CrossRef]
6. Cameron, E.; Campbell, A.; Jack, T. The orthomolecular treatment of cancer. III. Reticulum cell sarcoma: Double complete regression induced by high-dose ascorbic acid therapy. *Chem. Biol. Interact.* **1975**, *11*, 387–393. [CrossRef]
7. Cameron, E.; Pauling, L. Supplemental ascorbate in the supportive treatment of cancer: Prolongation of survival times in terminal human cancer. *Proc. Natl. Acad. Sci. USA* **1976**, *73*, 3685–3689. [CrossRef] [PubMed]
8. Cameron, E.; Pauling, L. Supplemental ascorbate in the supportive treatment of cancer: Reevaluation of prolongation of survival times in terminal human cancer. *Proc. Natl. Acad. Sci. USA* **1978**, *75*, 4538–4542. [CrossRef] [PubMed]
9. Padayatty, S.J.; Levine, M. Reevaluation of ascorbate in cancer treatment: Emerging evidence, open minds and serendipity. *J. Am. Coll. Nutr.* **2000**, *19*, 423–425. [CrossRef] [PubMed]
10. Creagan, E.T.; Moertel, C.G.; O'Fallon, J.R.; Schutt, A.J.; O'Connell, M.J.; Rubin, J.; Frytak, S. Failure of high-dose vitamin C (ascorbic acid) therapy to benefit patients with advanced cancer. A controlled trial. *N. Engl. J. Med.* **1979**, *301*, 687–690. [CrossRef] [PubMed]

11. Jacobs, C.; Hutton, B.; Ng, T.; Shorr, R.; Clemons, M. Is there a role for oral or intravenous ascorbate (vitamin C) in treating patients with cancer? A systematic review. *Oncologist* **2015**, *20*, 210–223. [CrossRef] [PubMed]

12. Wittes, R.E. Vitamin C and cancer. *N. Engl. J. Med.* **1985**, *312*, 178–179. [CrossRef] [PubMed]

13. Levine, M.; Conry-Cantilena, C.; Wang, Y.; Welch, R.W.; Washko, P.W.; Dhariwal, K.R.; Park, J.B.; Lazarev, A.; Graumlich, J.F.; King, J.; et al. Vitamin C pharmacokinetics in healthy volunteers: Evidence for a recommended dietary allowance. *Proc. Natl. Acad. Sci. USA* **1996**, *93*, 3704–3709. [CrossRef] [PubMed]

14. Levine, M.; Wang, Y.; Padayatty, S.J.; Morrow, J. A new recommended dietary allowance of vitamin C for healthy young women. *Proc. Natl. Acad. Sci. USA* **2001**, *98*, 9842–9846. [CrossRef] [PubMed]

15. Du, J.; Cullen, J.J.; Buettner, G.R. Ascorbic acid: Chemistry, biology and the treatment of cancer. *Biochim. Biophys. Acta* **2012**, *1826*, 443–457. [CrossRef] [PubMed]

16. Padayatty, S.J.; Sun, H.; Wang, Y.; Riordan, H.D.; Hewitt, S.M.; Katz, A.; Wesley, R.A.; Levine, M. Vitamin C pharmacokinetics: Implications for oral and intravenous use. *Ann. Intern. Med.* **2004**, *140*, 533–537. [CrossRef] [PubMed]

17. Chen, Q.; Espey, M.G.; Sun, A.Y.; Pooput, C.; Kirk, K.L.; Krishna, M.C.; Khosh, D.B.; Drisko, J.; Levine, M. Pharmacologic doses of ascorbate act as a prooxidant and decrease growth of aggressive tumor xenografts in mice. *Proc. Natl. Acad. Sci. USA* **2008**, *105*, 11105–11109. [CrossRef] [PubMed]

18. Hoffer, L.J.; Levine, M.; Assouline, S.; Melnychuk, D.; Padayatty, S.J.; Rosadiuk, K.; Rousseau, C.; Robitaille, L.; Miller, W.H., Jr. Phase I clinical trial of i.v. ascorbic acid in advanced malignancy. *Ann. Oncol.* **2008**, *19*, 1969–1974. [CrossRef] [PubMed]

19. Graumlich, J.F.; Ludden, T.M.; Conry-Cantilena, C.; Cantilena, L.R., Jr.; Wang, Y.; Levine, M. Pharmacokinetic model of ascorbic acid in healthy male volunteers during depletion and repletion. *Pharm. Res.* **1997**, *14*, 1133–1139. [CrossRef] [PubMed]

20. Ohno, S.; Ohno, Y.; Suzuki, N.; Soma, G.; Inoue, M. High-dose vitamin C (ascorbic acid) therapy in the treatment of patients with advanced cancer. *Anticancer Res.* **2009**, *29*, 809–815. [PubMed]

21. Monti, D.A.; Mitchell, E.; Bazzan, A.J.; Littman, S.; Zabrecky, G.; Yeo, C.J.; Pillai, M.V.; Newberg, A.B.; Deshmukh, S.; Levine, M. Phase I evaluation of intravenous ascorbic acid in combination with gemcitabine and erlotinib in patients with metastatic pancreatic cancer. *PLoS ONE* **2012**, *7*, e29794. [CrossRef] [PubMed]

22. Hoffer, L.J.; Robitaille, L.; Zakarian, R.; Melnychuk, D.; Kavan, P.; Agulnik, J.; Cohen, V.; Small, D.; Miller, W.H., Jr. High-dose intravenous vitamin C combined with cytotoxic chemotherapy in patients with advanced cancer: A phase I-II clinical trial. *PLoS ONE* **2015**, *10*, e0120228. [CrossRef] [PubMed]

23. Riordan, H.D.; Casciari, J.J.; Gonzalez, M.J.; Riordan, N.H.; Miranda-Massari, J.R.; Taylor, P.; Jackson, J.A. A pilot clinical study of continuous intravenous ascorbate in terminal cancer patients. *P. R. Health Sci. J.* **2005**, *24*, 269–276. [PubMed]

24. Duconge, J.; Miranda-Massari, J.R.; Gonzalez, M.J.; Jackson, J.A.; Warnock, W.; Riordan, N.H. Pharmacokinetics of vitamin C: Insights into the oral and intravenous administration of ascorbate. *P. R. Health Sci. J.* **2008**, *27*, 7–19. [PubMed]

25. Pauling, L. Are recommended daily allowances for vitamin C adequate? *Proc. Natl. Acad. Sci. USA* **1974**, *71*, 4442–4446. [CrossRef] [PubMed]

26. Langemann, H.; Torhorst, J.; Kabiersch, A.; Krenger, W.; Honegger, C.G. Quantitative determination of water- and lipid-soluble antioxidants in neoplastic and non-neoplastic human breast tissue. *Int. J. Cancer* **1989**, *43*, 1169–1173. [CrossRef] [PubMed]

27. Honegger, C.G.; Torhorst, J.; Langemann, H.; Kabiersch, A.; Krenger, W. Quantitative determination of water-soluble scavengers in neoplastic and non-neoplastic human breast tissue. *Int. J. Cancer* **1988**, *41*, 690–694. [CrossRef] [PubMed]

28. Agus, D.B.; Vera, J.C.; Golde, D.W. Stromal cell oxidation: A mechanism by which tumors obtain vitamin C. *Cancer Res.* **1999**, *59*, 4555–4558. [PubMed]

29. Violet, P.C.; Levine, M. Pharmacologic Ascorbate in Myeloma Treatment: Doses Matter. *EBioMedicine* **2017**, *18*, 9–10. [CrossRef] [PubMed]

30. Schoenfeld, J.D.; Sibenaller, Z.A.; Mapuskar, K.A.; Wagner, B.A.; Cramer-Morales, K.L.; Furqan, M.; Sandhu, S.; Carlisle, T.L.; Smith, M.C.; Abu Hejleh, T.; et al. O^{2-} and H_2O_2-Mediated Disruption of Fe Metabolism Causes the Differential Susceptibility of NSCLC and GBM Cancer Cells to Pharmacological Ascorbate. *Cancer Cell* **2017**, *31*, 487–500. [CrossRef] [PubMed]

31. Antunes, F.; Cadenas, E. Estimation of H_2O_2 gradients across biomembranes. *FEBS Lett.* **2000**, *475*, 121–126. [CrossRef]
32. Hyslop, P.A.; Hinshaw, D.B.; Halsey, W.A., Jr.; Schraufstatter, I.U.; Sauerheber, R.D.; Spragg, R.G.; Jackson, J.H.; Cochrane, C.G. Mechanisms of oxidant-mediated cell injury. The glycolytic and mitochondrial pathways of ADP phosphorylation are major intracellular targets inactivated by hydrogen peroxide. *J. Biol. Chem.* **1988**, *263*, 1665–1675. [PubMed]
33. Ahmad, I.M.; Aykin-Burns, N.; Sim, J.E.; Walsh, S.A.; Higashikubo, R.; Buettner, G.R.; Venkataraman, S.; Mackey, M.A.; Flanagan, S.W.; Oberley, L.W.; et al. Mitochondrial O_2^{-} and H_2O_2 mediate glucose deprivation-induced stress in human cancer cells. *J. Biol. Chem.* **2005**, *280*, 4254–4263. [CrossRef] [PubMed]
34. Yun, J.; Mullarky, E.; Lu, C.; Bosch, K.N.; Kavalier, A.; Rivera, K.; Roper, J.; Chio, I.I.; Giannopoulou, E.G.; Rago, C.; et al. Vitamin C selectively kills KRAS and BRAF mutant colorectal cancer cells by targeting GAPDH. *Science* **2015**, *350*, 1391–1396. [CrossRef] [PubMed]
35. Van der Reest, J.; Gottlieb, E. Anti-cancer effects of vitamin C. revisited. *Cell Res.* **2016**, *26*, 269–270. [CrossRef] [PubMed]
36. Chen, Q.; Espey, M.G.; Krishna, M.C.; Mitchell, J.B.; Corpe, C.P.; Buettner, G.R.; Shacter, E.; Levine, M. Pharmacologic ascorbic acid concentrations selectively kill cancer cells: Action as a pro-drug to deliver hydrogen peroxide to tissues. *Proc. Natl. Acad. Sci. USA* **2005**, *102*, 13604–13609. [CrossRef] [PubMed]
37. Chen, S.; Roffey, D.M.; Dion, C.A.; Arab, A.; Wai, E.K. Effect of Perioperative Vitamin C Supplementation on Postoperative Pain and the Incidence of Chronic Regional Pain Syndrome: A Systematic Review and Meta-Analysis. *Clin. J. Pain* **2016**, *32*, 179–185. [CrossRef] [PubMed]
38. Blaschke, K.; Ebata, K.T.; Karimi, M.M.; Zepeda-Martinez, J.A.; Goyal, P.; Mahapatra, S.; Tam, A.; Laird, D.J.; Hirst, M.; Rao, A.; et al. Vitamin C induces Tet-dependent DNA demethylation and a blastocyst-like state in ES cells. *Nature* **2013**, *500*, 222–226. [CrossRef] [PubMed]
39. Minor, E.A.; Court, B.L.; Young, J.I.; Wang, G. Ascorbate induces ten-eleven translocation (Tet) methylcytosine dioxygenase-mediated generation of 5-hydroxymethylcytosine. *J. Biol. Chem.* **2013**, *288*, 13669–13674. [CrossRef] [PubMed]
40. Shenoy, N.; Bhagat, T.; Nieves, E.; Stenson, M.; Lawson, J.; Choudhary, G.S.; Habermann, T.; Nowakowski, G.; Singh, R.; Wu, X.; et al. Upregulation of TET activity with ascorbic acid induces epigenetic modulation of lymphoma cells. *Blood Cancer J.* **2017**, *7*, e587. [CrossRef] [PubMed]
41. Wu, X.; Zhang, Y. TET-mediated active DNA demethylation: Mechanism, function and beyond. *Nat. Rev. Genet.* **2017**, *18*, 517–534. [CrossRef] [PubMed]
42. Agathocleous, M.; Meacham, C.E.; Burgess, R.J.; Piskounova, E.; Zhao, Z.; Crane, G.M.; Cowin, B.L.; Bruner, E.; Murphy, M.M.; Chen, W.; et al. Ascorbate regulates haematopoietic stem cell function and leukaemogenesis. *Nature* **2017**, *549*, 476–481. [CrossRef] [PubMed]
43. Masoud, G.N.; Li, W. HIF-1alpha pathway: Role, regulation and intervention for cancer therapy. *Acta Pharm. Sin. B* **2015**, *5*, 378–389. [CrossRef] [PubMed]
44. Campbell, E.J.; Vissers, M.C.; Bozonet, S.; Dyer, A.; Robinson, B.A.; Dachs, G.U. Restoring physiological levels of ascorbate slows tumor growth and moderates HIF-1 pathway activity in $Gulo^{-/-}$ mice. *Cancer Med.* **2015**, *4*, 303–314. [CrossRef] [PubMed]
45. Carr, A.C.; Vissers, M.C.; Cook, J.S. The effect of intravenous vitamin C on cancer- and chemotherapy-related fatigue and quality of life. *Front. Oncol.* **2014**, *4*, 283. [CrossRef] [PubMed]
46. Chen, Q.; Espey, M.G.; Sun, A.Y.; Lee, J.H.; Krishna, M.C.; Shacter, E.; Choyke, P.L.; Pooput, C.; Kirk, K.L.; Buettner, G.R.; et al. Ascorbate in pharmacologic concentrations selectively generates ascorbate radical and hydrogen peroxide in extracellular fluid in vivo. *Proc. Natl. Acad. Sci. USA* **2007**, *104*, 8749–8754. [CrossRef] [PubMed]
47. Chen, P.; Yu, J.; Chalmers, B.; Drisko, J.; Yang, J.; Li, B.; Chen, Q. Pharmacological ascorbate induces cytotoxicity in prostate cancer cells through ATP depletion and induction of autophagy. *Anticancer Drugs* **2012**, *23*, 437–444. [CrossRef] [PubMed]
48. Espey, M.G.; Chen, P.; Chalmers, B.; Drisko, J.; Sun, A.Y.; Levine, M.; Chen, Q. Pharmacologic ascorbate synergizes with gemcitabine in preclinical models of pancreatic cancer. *Free Radic. Biol. Med.* **2011**, *50*, 1610–1619. [CrossRef] [PubMed]
49. Ong, C.P. High Dose Vitamin C and Low Dose Chemo Treatment. *J. Cancer Sci.* **2018**, *5*, 4.

50. Ma, Y.; Chapman, J.; Levine, M.; Polireddy, K.; Drisko, J.; Chen, Q. High-dose parenteral ascorbate enhanced chemosensitivity of ovarian cancer and reduced toxicity of chemotherapy. *Sci. Transl. Med.* **2014**, *6*, 222ra18. [CrossRef] [PubMed]
51. Welsh, J.L.; Wagner, B.A.; van't Erve, T.J.; Zehr, P.S.; Berg, D.J.; Halfdanarson, T.R.; Yee, N.S.; Bodeker, K.L.; Du, J.; Roberts, L.J., 2nd; et al. Pharmacological ascorbate with gemcitabine for the control of metastatic and node-positive pancreatic cancer (PACMAN): Results from a phase I. clinical trial. *Cancer Chemother. Pharmacol.* **2013**, *71*, 765–775. [CrossRef] [PubMed]
52. Berenson, J.R.; Matous, J.; Swift, R.A.; Mapes, R.; Morrison, B.; Yeh, H.S. A phase I/II study of arsenic trioxide/bortezomib/ascorbic acid combination therapy for the treatment of relapsed or refractory multiple myeloma. *Clin. Cancer Res.* **2007**, *13*, 1762–1768. [CrossRef] [PubMed]
53. Berenson, J.R.; Boccia, R.; Siegel, D.; Bozdech, M.; Bessudo, A.; Stadtmauer, E.; Talisman Pomeroy, J.; Steis, R.; Flam, M.; Lutzky, J.; et al. Efficacy and safety of melphalan, arsenic trioxide and ascorbic acid combination therapy in patients with relapsed or refractory multiple myeloma: A prospective, multicentre, phase II, single-arm study. *Br. J. Haematol.* **2006**, *135*, 174–183. [CrossRef] [PubMed]
54. Abou-Jawde, R.M.; Reed, J.; Kelly, M.; Walker, E.; Andresen, S.; Baz, R.; Karam, M.A.; Hussein, M. Efficacy and safety results with the combination therapy of arsenic trioxide, dexamethasone, and ascorbic acid in multiple myeloma patients: A phase 2 trial. *Med. Oncol.* **2006**, *23*, 263–272. [CrossRef] [PubMed]
55. Chang, J.E.; Voorhees, P.M.; Kolesar, J.M.; Ahuja, H.G.; Sanchez, F.A.; Rodriguez, G.A.; Kim, K.; Werndli, J.; Bailey, H.H.; Kahl, B.S. Phase II study of arsenic trioxide and ascorbic acid for relapsed or refractory lymphoid malignancies: A Wisconsin Oncology Network study. *Hematol. Oncol.* **2009**, *27*, 11–16. [CrossRef] [PubMed]
56. Bael, T.E.; Peterson, B.L.; Gollob, J.A. Phase II trial of arsenic trioxide and ascorbic acid with temozolomide in patients with metastatic melanoma with or without central nervous system metastases. *Melanoma Res.* **2008**, *18*, 147–151. [CrossRef] [PubMed]
57. Subbarayan, P.R.; Lima, M.; Ardalan, B. Arsenic trioxide/ascorbic acid therapy in patients with refractory metastatic colorectal carcinoma: A clinical experience. *Acta Oncol.* **2007**, *46*, 557–561. [CrossRef] [PubMed]
58. Wu, K.L.; Beksac, M.; van Droogenbroeck, J.; Amadori, S.; Zweegman, S.; Sonneveld, P. Phase II multicenter study of arsenic trioxide, ascorbic acid and dexamethasone in patients with relapsed or refractory multiple myeloma. *Haematologica* **2006**, *91*, 1722–1723. [PubMed]
59. Aldoss, I.; Mark, L.; Vrona, J.; Ramezani, L.; Weitz, I.; Mohrbacher, A.M.; Douer, D. Adding ascorbic acid to arsenic trioxide produces limited benefit in patients with acute myeloid leukemia excluding acute promyelocytic leukemia. *Ann. Hematol.* **2014**, *93*, 1839–1843. [CrossRef] [PubMed]
60. Bahlis, N.J.; McCafferty-Grad, J.; Jordan-McMurry, I.; Neil, J.; Reis, I.; Kharfan-Dabaja, M.; Eckman, J.; Goodman, M.; Fernandez, H.F.; Boise, L.H.; et al. Feasibility and correlates of arsenic trioxide combined with ascorbic acid-mediated depletion of intracellular glutathione for the treatment of relapsed/refractory multiple myeloma. *Clin. Cancer Res.* **2002**, *8*, 3658–3668. [PubMed]
61. Held, L.A.; Rizzieri, D.; Long, G.D.; Gockerman, J.P.; Diehl, L.F.; de Castro, C.M.; Moore, J.O.; Horwitz, M.E.; Chao, N.J.; Gasparetto, C. A Phase I study of arsenic trioxide (Trisenox), ascorbic acid, and bortezomib (Velcade) combination therapy in patients with relapsed/refractory multiple myeloma. *Cancer Investig.* **2013**, *31*, 172–176. [CrossRef] [PubMed]
62. Welch, J.S.; Klco, J.M.; Gao, F.; Procknow, E.; Uy, G.L.; Stockerl-Goldstein, K.E.; Abboud, C.N.; Westervelt, P.; DiPersio, J.F.; Hassan, A.; et al. Combination decitabine, arsenic trioxide, and ascorbic acid for the treatment of myelodysplastic syndrome and acute myeloid leukemia: A phase I study. *Am. J. Hematol.* **2011**, *86*, 796–800. [CrossRef] [PubMed]
63. Stephenson, C.M.; Levin, R.D.; Spector, T.; Lis, C.G. Phase I clinical trial to evaluate the safety, tolerability, and pharmacokinetics of high-dose intravenous ascorbic acid in patients with advanced cancer. *Cancer Chemother. Pharmacol.* **2013**, *72*, 139–146. [CrossRef] [PubMed]
64. Kawada, H.; Sawanobori, M.; Tsuma-Kaneko, M.; Wasada, I.; Miyamoto, M.; Murayama, H.; Toyosaki, M.; Onizuka, M.; Tsuboi, K.; Tazume, K.; et al. Phase I Clinical Trial of Intravenous L-ascorbic Acid Following Salvage Chemotherapy for Relapsed B-cell non-Hodgkin's Lymphoma. *Tokai J. Exp. Clin. Med.* **2014**, *39*, 111–115. [PubMed]
65. Polireddy, K.; Dong, R.; Reed, G.; Yu, J.; Chen, P.; Williamson, S.; Violet, P.C.; Pessetto, Z.; Godwin, A.K.; Fan, F.; et al. High Dose Parenteral Ascorbate Inhibited Pancreatic Cancer Growth and Metastasis: Mechanisms and a Phase, I./IIa study. *Sci. Rep.* **2017**, *7*, 17188. [CrossRef] [PubMed]

66. Nielsen, T.K.; Hojgaard, M.; Andersen, J.T.; Jorgensen, N.R.; Zerahn, B.; Kristensen, B.; Henriksen, T.; Lykkesfeldt, J.; Mikines, K.J.; Poulsen, H.E. Weekly ascorbic acid infusion in castration-resistant prostate cancer patients: A single-arm phase II trial. *Transl. Androl. Urol.* **2017**, *6*, 517–528. [CrossRef] [PubMed]

67. Mosteller, R.D. Simplified calculation of body-surface area. *N. Engl. J. Med.* **1987**, *317*, 1098. [CrossRef] [PubMed]

68. Vollbracht, C.; Schneider, B.; Leendert, V.; Weiss, G.; Auerbach, L.; Beuth, J. Intravenous vitamin C administration improves quality of life in breast cancer patients during chemo-/radiotherapy and aftercare: Results of a retrospective, multicentre, epidemiological cohort study in Germany. *In Vivo* **2011**, *25*, 983–990. [PubMed]

69. Riordan, H.D.; Hunninghake, R.B.; Riordan, N.H.; Jackson, J.J.; Meng, X.; Taylor, P.; Casciari, J.J.; Gonzalez, M.J.; Miranda-Massari, J.R.; Mora, E.M.; et al. Intravenous ascorbic acid: Protocol for its application and use. *P. R. Health Sci. J.* **2003**, *22*, 287–290. [PubMed]

70. Padayatty, S.J.; Riordan, H.D.; Hewitt, S.M.; Katz, A.; Hoffer, L.J.; Levine, M. Intravenously administered vitamin C as cancer therapy: Three cases. *CMAJ* **2006**, *174*, 937–942. [CrossRef] [PubMed]

antioxidants

MDPI

Case Report

Intravenous Vitamin C Administration Improved Blood Cell Counts and Health-Related Quality of Life of Patient with History of Relapsed Acute Myeloid Leukaemia

Mike N. Foster [1],*, Anitra C. Carr [2], Alina Antony [3], Selene Peng [3] and Mike G. Fitzpatrick [3]

[1]   Integrated Health Options Ltd., Auckland 1050, New Zealand
[2]   Department of Pathology and Biomedical Science, University of Otago, Christchurch 8011, New Zealand; anitra.carr@otago.ac.nz
[3]   Feedback Research Ltd., Auckland 1050, New Zealand; alina@feedbackresearch.co.nz (A.A.); selene@feedbackresearch.co.nz (S.P.); mike@feedbackresearch.co.nz (M.G.F.)
*   Correspondence: mfoster@integratedhealthoptions.co.nz; Tel.: +64-9524-7745

Received: 8 June 2018; Accepted: 12 July 2018; Published: 16 July 2018

Abstract: A 52-year-old female presented to Integrated Health Options Clinic in October 2014 with a history of relapsed acute myeloid leukaemia (AML, diagnosed in 2009 and relapsed in 2014). Intravenous(IV) vitamin C therapy was initiated (in 2014) following completion of chemotherapy as an alternative to haematopoietic stem cell transplantation. IV vitamin C was administered twice weekly at a dose of 70 g/infusion. Within 4 weeks of initiation of IV vitamin C therapy, there was a dramatic improvement in the patient's blood indices with platelet cell counts increasing from 25×10^9/L to 196×10^9/L and white blood cell counts increasing from 0.29×10^9/L to 4.0×10^9/L, with further improvements observed over the next 18 months. Furthermore, there was a clear and sustained improvement in the patient's health-related quality of life scores assessed using a validated questionnaire. She has remained healthy and in complete remission until the present day. This case study highlights the benefits of IV vitamin C as a supportive therapy for previously relapsed AML.

Keywords: vitamin C; intravenous vitamin C; leukemia; acute myeloid leukemia; AML

1. Introduction

Up to 70% of acute myeloid leukaemia (AML) patients treated with chemotherapy will relapse and overall five year survival is about 25% [1,2]. Treatment options for relapsed AML are currently limited to haematopoietic stem cell (HSC) transplantation [3]. Two recent studies have indicated a role for vitamin C in suppressing leukaemia progression through epigenetic mechanisms [4,5]. Vitamin C was found to enhance HSC differentiation and decrease the number of circulating blast cells and leukaemia progression in animal models resulting in improved survival. Animals with vitamin C deficiency were found to have higher levels of HSCs with enhanced self-renewal function, which increases susceptibility to leukaemia. Patients with haematological malignances have lower vitamin C status than control subjects, with up to 58% exhibiting deficiency [6,7], and further depletion of vitamin C can occur following chemotherapy and HSC transplantation [8,9]. Thus, administration of vitamin C to deficient patients may help to restore normal HSC function and differentiation. Herein we report on a case of relapsed AML who underwent regular IV vitamin C administration instead of HSC transplantation.

2. Case Report

The patient signed a consent form allowing the presentation of her case. On 5 May 2009, a 47-year-old female patient was admitted to Auckland Hospital, New Zealand, and diagnosed with Acute Myeloid Leukaemia (AML, bone marrow 75% blasts, normal karyotype, NPM+, FLT3+ with allelic ratio 0.01). She was prescribed daunorubicin/cytarabine (ARA-C) induction, and a Groshong line was inserted. She underwent four cycles of chemotherapy with DA (daunorubicin, cytarabine) × 2MACE (amsacrine, cytarabine, etoposide) and MidAC (mitoxantrone, cytarabine). This was complicated by a presumed Aspergillus invasive pulmonary infection and was treated with oral Voriconazole for three months. A goitre was noted on admission and she was diagnosed with hypothyroidism (thyroid-stimulating hormone (TSH) 54 mU/L, thyroxine (T4) 7.7 pmol/L). She had positive thyroid antibodies and was prescribed thyroxine 50 mcg/day.

Following every cycle of chemotherapy, she was readmitted with a neutropenic fever, which required antibiotic therapy. On 9 May 2009, she developed a neutropenic fever and was started on empiric cefepime and gentamycin while awaiting blood culture results. On those antibiotics, she developed diarrhoea and a rash on her legs, which was diagnosed as Erythema multiforme and thought to be caused by either the cefepime or allopurinol, which she had also been prescribed. Both drugs were discontinued, and she was trialled on meropenem. This drug is a beta-lactam antibiotic given by IV for more serious infections.

By 14 May 2009, her fever persisted and no obvious cause was found at that time. High-resolution computed tomography (HRCT) confirmed that the chest was normal. IV Amphotericin B was commenced. Finally, Staphylococcus epidermidis was grown from a blood culture sample taken from the Groshong line. Amphotericin B was discontinued and IV vancomycin commenced. Her fever was resolved and she was discharged home with IV vancomycin for a further five days. Her platelet count prior to discharge was <10 × 10^9/L. She was given a platelet transfusion prior to her discharge.

On 25 May 2009, she was readmitted with neutropenic fever. She was again treated with meropenem and vancomycin. She had haemoglobin (Hb) 88 g/L and platelets 20 × 10^9/L on admission. She was transfused with red blood cells and platelets but developed a reaction to the platelet transfusion, which was managed with hydrocortisone and cetirizine. After two days she was discharged on augmentin for five days and a further two days of vancomycin. The four cycles of chemotherapy were completed by October 2009.

She underwent regular review by the Hematology Clinic until 5 May 2014 when she was discharged from the clinic in complete remission. She remained healthy until August 2014 when she again developed a neutropenia and thrombocytopenia. A bone marrow aspirate on 20 August 2014 revealed a relapsed AML with 61% blasts of similar morphology and cell marker expression as the initial diagnosis. At this time she also had borderline glucose intolerance with HbA1c of 40 mmol/mol. She saw a haematologist, who suggested more intensive chemotherapy, and allogeneic bone marrow transplant. The Haematologist stated, "without a transplant, cure is extremely unlikely and even with a transplant, she is faced with significant transplant related morbidity and an appreciable relapse risk, but with a potential cure opportunity".

The patient decided to undergo further chemotherapy but rejected a bone marrow transplant. Induction chemotherapy was commenced on 29 August 2014. Bone marrow biopsy on 1 October 2014, showed complete remission of her AML (with 4% blasts). She had no further chemotherapy.

Intravenous Vitamin C Therapy

On the 10 October 2014, the patient had her first consultation at Integrated Health Options clinic, Auckland, New Zealand. Her Hb at that visit was 93 g/L and platelets 27 × 10^9/L. There was a discussion with the patient, in the presence of her daughter, about the clinic's protocol that IV vitamin C infusions are not routinely offered to people with a platelet count of less than 50 × 10^9/L. After informed consent, she chose to have the IV vitamin C infusions. Her G6PD was normal (tested due to potential haemolytic anaemia with G6PD deficiency). She commenced IV vitamin C therapy on the 7 November 2014 and was prescribed twice weekly IV vitamin C infusions at 70 g/infusion. Her initial pre-infusion plasma vitamin C concentration, measured prior to her first test dose of IV vitamin C on 7 November 2014, was 0.66 mg/dL (37.5 µmol/L). This plasma concentration is considered inadequate, with ≥50 µmol/L (0.88 mg/dL) being adequate and ≥70 µmol/L (1.23 mg/dL) being saturating [10,11]. Post-infusion plasma vitamin C levels were regularly assessed to confirm that the proposed ascorbic acid therapeutic level was achieved (Table 1) [12]. The patient underwent weekly complete blood counts (CBC) and renal function tests, and was also prescribed additional oral supplements (Table 2).

Table 1. Intravenous vitamin C dosage infused into the patient and post-infusion ascorbic acid therapeutic level in blood plasma.

Date (Day/Month/Year)	IV Vitamin C Dose (g)	AATL (mg/dL) [1]	AATL (mmol/L)
21/11/14	70	451	26
8/01/15	70	286	16
9/01/15	75	435	25
17/02/15	70	373	22
24/03/15	70	394	22
8/05/15	70	465	26
12/06/15	65	373	21
17/08/15	65	350	20
15/09/15	65	347	20
1/10/15	65	368	21
3/12/15	65	369	21
9/02/16	65	315	18
19/02/16	65	389	22
24/06/16	65	352	20

[1] Ascorbic acid was measured using high performance liquid chromatography (HPLC) with electrochemical detection [13]. Targeted level for ascorbic acid therapeutic level (AATL) is 350–400 mg/dL (or 20–23 mmol/L) [14,15].

Table 2. Supplementary medication taken concurrently with IV vitamin C infusion by patient.

Supplements	Intake/Day
Alpha lipoic acid [1]	600 mg
Sodium ascorbate	2000 mg
Methylated B vitamins [2]	1000 mg
Vitamin D	5000 units
Vitamin K1	1500 mcg
Vitamin K2-MK4	1000 mcg
Vitamin K2-MK7	200 mcg

[1] Mixture of R and S enantiomers. [2] The biologically active form of multiple B vitamins.

Prior to initiation of IV vitamin C treatments, the patient's Hb was 107 g/L, platelets 25×10^9/L and white blood cells (WBC) 0.29×10^9/L. By the 19 December 2014, 42 days after beginning IV vitamin C infusions, her Hb had increased to 111 g/L, platelets to 196×10^9/L (Figure 1) and WBC to 4×10^9/L (Figure 2). On the 7 September 2017 her Hb was 124 g/L, platelet counts were 227×10^9/L, and WBC count was 6.5×10^9/L with a normal differential.

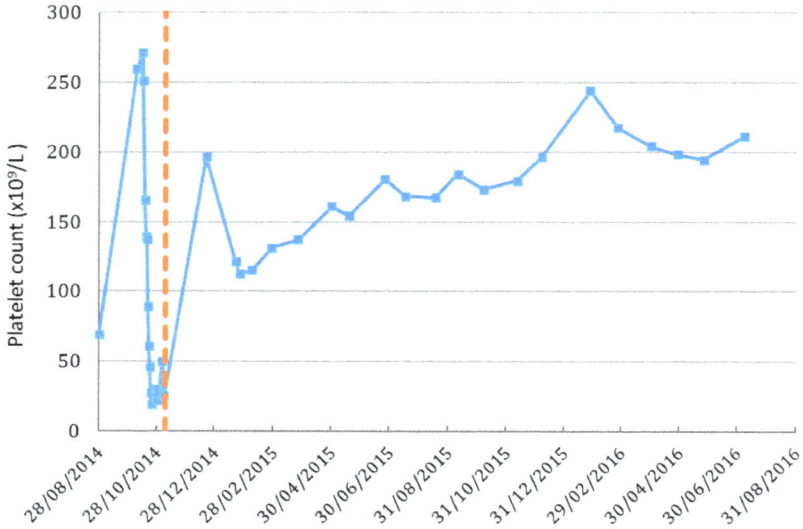

Figure 1. Platelet cell count in patient before and after IV vitamin C treatment. Dashed line indicates the IV vitamin C commenced date. Normal range of platelet count is 150–400 ($\times 10^9$/L).

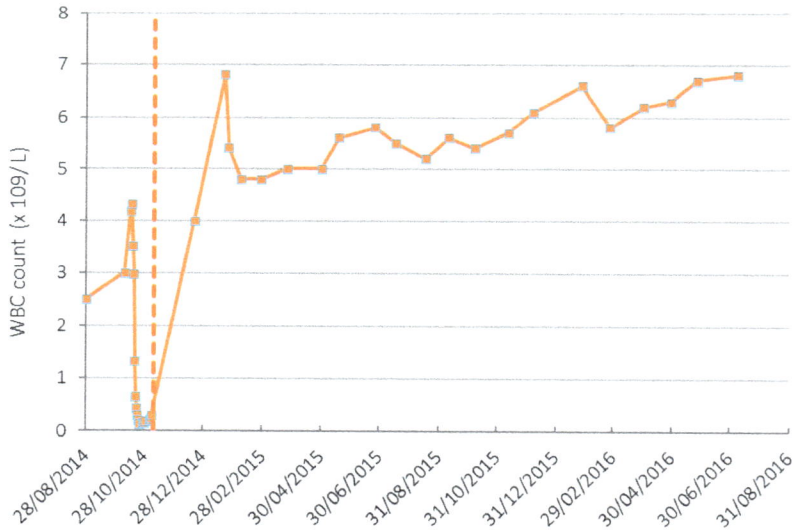

Figure 2. White blood cell (WBC) count in patient before and after IV vitamin C treatment. Dashed line indicates the IV vitamin C commenced date. Normal range of WBC count is 4.5–11 ($\times 10^9$/L).

The quality of life (QoL) of the patient was measured before and throughout the IV vitamin C therapy using the validated European Organisation for Research and Treatment of Cancer Quality of Life Questionnaire (EORTC QLQ-C30), which has been used previously to assess QoL following IV vitamin C therapy [16,17]. Six weeks after starting IV vitamin C infusions, her global health status score had increased from 58 to 83, her overall functional score had increased from 89 to 98, and her overall symptom score had dropped from 39 to 8 (Table 3). This was a marked improvement of her QoL, which has been maintained until the present time.

Table 3. Quality of life score measured using the EORTC QLQ-C30.

Date (Day/Month/Year)	Global Health Status	Functional Scales	Symptom Scales
Baseline (10/10/14) [1]	58.3	88.9	38.5
6 weeks (17/12/14)	83.3	97.6	7.7
3 months (05/02/15)	83.3	97.8	5.1
6 months (07/05/15)	83.3	95.6	7.7
12 months (11/11/15)	83.3	97.8	7.7

[1] Baseline survey was completed at first consultation. The following survey rounds were calculated from the first IV vitamin C treatment date—7/11/14.

The patient continued twice weekly IV vitamin C infusions of 70 g until August 2015, when the patient felt so well that she decided to reduce the frequency of her IV vitamin C infusions to once weekly. As of 21 September 2017, she has had 141 IV vitamin C infusions. Her blood vitamin C level (AATL) post-infusion averaged at 360 mg/dL (targeted therapeutic range is 350–400 mg/dL) [14,15]. She has now cut back to a monthly infusion but maintains her oral intake of vitamin C as well as the other supplements. The patient remains clinically very well and has remained in complete remission.

3. Discussion

The use of complementary and alternative medicine is substantial in the general population [18]. High-dose IV vitamin C, in particular, is sought after by many cancer patients [19]. The rationale for IV administration is the ability to bypass the regulated intestinal uptake of oral vitamin C, thus resulting in 20-fold higher plasma concentrations with IV infusion [20]. These high concentrations of vitamin C are able to indirectly generate hydrogen peroxide through a transition metal ion-dependent mechanism, and this is thought to specifically target and kill cancer cells [21]. IV administration may also be required to facilitate diffusion of vitamin C into tumours with subsequent modulation of important cell signalling pathways [22,23]. Although evidence for the efficacy of vitamin C against solid tumours is currently limited, recent research is indicating a possible role for vitamin C in modulating certain haematological malignancies that are sensitive to the epigenetic regulating effects of vitamin C [4,5]. Vitamin C is able to upregulate the activity of specific tumour suppressor epigenetic enzymes (TETs), which contain mutations in a proportion of AML patients [24,25].

The purpose of this case study was to document the benefits of IV vitamin C as a supportive therapy for previously relapsed AML. In this case, after IV vitamin C treatment, the patient regained normal values for absolute circulating neutrophil count ($>1 \times 10^9$/L) and platelets count ($>100 \times 10^9$/L). Cell culture studies have indicated that vitamin C can restore normal tumour suppressor expression profiles and the maturation of HSCs [4,26]. Although the epigenetic (TET mutation) status of the patient was not assessed, it is possible that she had the mutations which respond to vitamin C treatment. The patient remains clinically very well. In addition to vitamin supplements, the patient also incorporated a healthy lifestyle and dietary changes, such as vegetarian diet, yoga, meditation, homeopathic remedies, and herbal remedies (e.g., mistletoe, and black cumin seed oil). These supplements and lifestyle changes may have contributed to the dramatic results achieved by this patient. Therefore, the relative contribution of IV vitamin C therapy could not be singled out. It is also possible that vitamin C may work synergistically with other supplements. Thus, in this case, it was observed that chemotherapy followed by IV vitamin C therapy, combined with a healthy diet and other supplements, supported ongoing patient quality of life and remission of AML.

Administration of chemotherapeutic drugs to patients can result in depletion of vitamin C levels and negatively impact on the patient's health-related quality of life [17]. Thus, administration of vitamin C to these patients may contribute to improved quality of life through replenishing the essential nutrient and counteracting the toxic side effects of chemotherapy. Indeed, the patient had an inadequate vitamin C status initially, and we observed a marked improvement in the patient's health-related quality of life following initiation of IV vitamin C therapy, which has been maintained until the present time. The incidence of AML increases dramatically with age and prognosis for the elderly is poor with only 10% surviving five years after diagnosis [2]. As a result of the toxic effects of therapy, including myelosuppression and an increased risk of infection, chemotherapy may not be offered to the very elderly, in which case IV vitamin C may be a viable palliative option for these patients due to its relative safety. Furthermore, a recent trial of low gram dose IV vitamin C administered with DCAG (decitabine, granulocyte colony stimulating factor, cytarabine, aclarubicin) in elderly acute myeloid leukaemia patients showed an increased incidence of complete remission after first induction and prolonged overall survival compared with DCAG alone [27].

4. Conclusions

Administration of high-dose IV vitamin C to previously relapsed AML patients may help to restore normal blood cell counts and function. Improved quality of life is also observed. Future clinical trials of IV vitamin C for haematological cancers should target patients with the relevant (TET) mutations, or endeavour to include sub-group analysis of patients with these mutations.

Author Contributions: M.N.F. treated the patient and collected samples, A.A., S.P. and M.G.F. analysed samples and data, M.N.F. and A.C.C. wrote the paper, which was edited by A.A., S.P. and M.G.F.

Funding: This research received no external funding.

Acknowledgments: We thank the patient for giving us permission to present her results. A.C.C. is supported by a Health Research Council of New Zealand Sir Charles Hercus Health Research Fellowship (#16/037).

Conflicts of Interest: The authors declare no conflict of interest.

References

1. Yin, J.A.; Grimwade, D. Minimal residual disease evaluation in acute myeloid leukaemia. *Lancet* **2002**, *360*, 160–162. [CrossRef]
2. Howlader, N.; Noone, A.M.; Krapcho, M.; Miller, D.; Bishop, K.; Kosary, C.L.; Yu, M.; Ruhl, J.; Tatalovich, Z.; Mariotto, A.; et al. *SEER Cancer Statistics Review, 1975–2014*; National Cancer Institute: Bethesda, MD, USA, 2016.
3. Appelbaum, F.R. Hematopoietic cell transplantation beyond first remission. *Leukemia* **2002**, *16*, 157–159. [CrossRef] [PubMed]
4. Cimmino, L.; Dolgalev, I.; Wang, Y.; Yoshimi, A.; Martin, G.H.; Wang, J.; Ng, V.; Xia, B.; Witkowski, M.T.; Mitchell-Flack, M.; et al. Restoration of TET2 function blocks aberrant self-renewal and leukemia progression. *Cell* **2017**, *170*, 1079–1095.e20. [CrossRef] [PubMed]
5. Agathocleous, M.; Meacham, C.E.; Burgess, R.J.; Piskounova, E.; Zhao, Z.; Crane, G.M.; Cowin, B.L.; Bruner, E.; Murphy, M.M.; Chen, W.; et al. Ascorbate regulates haematopoietic stem cell function and leukaemogenesis. *Nature* **2017**, *549*, 476–481. [CrossRef] [PubMed]
6. Huijskens, M.J.; Wodzig, W.K.; Walczak, M.; Germeraad, W.T.; Bos, G.M. Ascorbic acid serum levels are reduced in patients with hematological malignancies. *Results Immunol.* **2016**, *6*, 8–10. [CrossRef] [PubMed]
7. Liu, M.; Ohtani, H.; Zhou, W.; Orskov, A.D.; Charlet, J.; Zhang, Y.W.; Shen, H.; Baylin, S.B.; Liang, G.; Gronbaek, K.; et al. Vitamin C increases viral mimicry induced by 5-aza-2′-deoxycytidine. *Proc. Natl. Acad. Sci. USA* **2016**, *113*, 10238–10244. [CrossRef] [PubMed]
8. Nannya, Y.; Shinohara, A.; Ichikawa, M.; Kurokawa, M. Serial profile of vitamins and trace elements during the acute phase of allogeneic stem cell transplantation. *Biol. Blood Marrow Transplant.* **2014**, *20*, 430–434. [CrossRef] [PubMed]
9. Rasheed, M.; Roberts, C.H.; Gupta, G.; Fisher, B.J.; Leslie, K.; Simmons, G.L.; Wiedle, C.M.; McCarty, J.M.; Clark, W.B.; Chung, H.M.; et al. Low plasma vitamin C levels in patients undergoing stem cell transplantation (Abstract). *Biol. Blood Marrow Transplant.* **2017**, *23*, S286–S287. [CrossRef]

10. Lykkesfeldt, J.; Poulsen, H.E. Is vitamin C supplementation beneficial? Lessons learned from randomised controlled trials. *Br. J. Nutr.* **2010**, *103*, 1251–1259. [CrossRef] [PubMed]

11. European Food Safety Authority Panel on Dietetic Products Nutrition and Allergies. Scientific opinion on dietary reference values for vitamin C. *EFSA J.* **2013**, *11*, 3418.

12. Welsh, J.L.; Wagner, B.A.; van't Erve, T.J.; Zehr, P.S.; Berg, D.J.; Halfdanarson, T.R.; Yee, N.S.; Bodeker, K.L.; Du, J.; Roberts, L.J., 2nd; et al. Pharmacological ascorbate with gemcitabine for the control of metastatic and node-positive pancreatic cancer (PACMAN): Results from a phase I clinical trial. *Cancer Chemother. Pharmacol.* **2013**, *71*, 765–775. [CrossRef] [PubMed]

13. Lykkesfeldt, J. Measurement of ascorbic acid and dehydroascorbic acid in biological samples. *Curr. Protoc. Toxicol.* **2002**. [CrossRef]

14. Mikirova, N.A.; Casciari, J.J.; Hunninghake, R.E.; Riordan, N.H. Intravenous ascorbic acid protocol for cancer patients: Scientific rationale, pharmacology, and clinical experience. *Funct. Foods Health Dis.* **2013**, *3*, 344–366.

15. Chen, Q.; Espey, M.G.; Krishna, M.C.; Mitchell, J.B.; Corpe, C.P.; Buettner, G.R.; Shacter, E.; Levine, M. Pharmacologic ascorbic acid concentrations selectively kill cancer cells: Action as a pro-drug to deliver hydrogen peroxide to tissues. *Proc. Natl. Acad. Sci. USA* **2005**, *102*, 13604–13609. [CrossRef] [PubMed]

16. Aaronson, N.K.; Ahmedzai, S.; Bergman, B.; Bullinger, M.; Cull, A.; Duez, N.J.; Filiberti, A.; Flechtner, H.; Fleishman, S.B.; de Haes, J.C.; et al. The European Organization for Research and Treatment of Cancer QLQ-C30: A quality-of-life instrument for use in international clinical trials in oncology. *J. Natl. Cancer Inst.* **1993**, *85*, 365–376. [CrossRef] [PubMed]

17. Carr, A.C.; Vissers, M.C.M.; Cook, J.S. The effect of intravenous vitamin C on cancer- and chemotherapy-related fatigue and quality of life. *Front. Oncol.* **2014**, *4*, 283. [CrossRef] [PubMed]

18. Harris, P.; Rees, R. The prevalence of complementary and alternative medicine use among the general population: A systematic review of the literature. *Complement. Ther. Med.* **2000**, *8*, 88–96. [CrossRef] [PubMed]

19. Padayatty, S.J.; Sun, A.Y.; Chen, Q.; Espey, M.G.; Drisko, J.; Levine, M. Vitamin C: Intravenous use by complementary and alternative medicine practitioners and adverse effects. *PLoS ONE* **2010**, *5*, e11414. [CrossRef] [PubMed]

20. Padayatty, S.J.; Sun, H.; Wang, Y.; Riordan, H.D.; Hewitt, S.M.; Katz, A.; Wesley, R.A.; Levine, M. Vitamin C pharmacokinetics: Implications for oral and intravenous use. *Ann. Intern. Med.* **2004**, *140*, 533–537. [CrossRef] [PubMed]

21. Parrow, N.L.; Leshin, J.A.; Levine, M. Parenteral ascorbate as a cancer therapeutic: A reassessment based on pharmacokinetics. *Antioxid. Redox Signal.* **2013**, *19*, 2141–2156. [CrossRef] [PubMed]

22. Kuiper, C.; Vissers, M.C.; Hicks, K.O. Pharmacokinetic modeling of ascorbate diffusion through normal and tumor tissue. *Free Radic. Biol. Med.* **2014**, *77*, 340–352. [CrossRef] [PubMed]

23. Kuiper, C.; Vissers, M.C. Ascorbate as a co-factor for Fe- and 2-oxoglutarate dependent dioxygenases: Physiological activity in tumor growth and progression. *Front. Oncol.* **2014**, *4*, 359. [CrossRef] [PubMed]

24. Mastrangelo, D.; Pelosi, E.; Castelli, G.; Lo-Coco, F.; Testa, U. Mechanisms of anti-cancer effects of ascorbate: Cytotoxic activity and epigenetic modulation. *Blood Cells Mol. Dis.* **2018**, *69*, 57–64. [CrossRef] [PubMed]

25. Delhommeau, F.; Dupont, S.; Della Valle, V.; James, C.; Trannoy, S.; Masse, A.; Kosmider, O.; Le Couedic, J.P.; Robert, F.; Alberdi, A.; et al. Mutation in TET2 in myeloid cancers. *N. Engl. J. Med.* **2009**, *360*, 2289–2301. [CrossRef] [PubMed]

26. Huijskens, M.J.; Walczak, M.; Koller, N.; Briede, J.J.; Senden-Gijsbers, B.L.; Schnijderberg, M.C.; Bos, G.M.; Germeraad, W.T. Technical advance: Ascorbic acid induces development of double-positive T cells from human hematopoietic stem cells in the absence of stromal cells. *J. Leukoc. Biol.* **2014**, *96*, 1165–1175. [CrossRef] [PubMed]

27. Zhao, H.; Zhu, H.; Huang, J.; Zhu, Y.; Hong, M.; Zhu, H.; Zhang, J.; Li, S.; Yang, L.; Lian, Y.; et al. The synergy of vitamin C with decitabine activates TET2 in leukemic cells and significantly improves overall survival in elderly patients with acute myeloid leukemia. *Leuk. Res.* **2018**, *66*, 1–7. [CrossRef] [PubMed]

antioxidants

MDPI

Case Report

The Use of Intravenous Vitamin C as a Supportive Therapy for a Patient with Glioblastoma Multiforme

Nicola Baillie [1,*], Anitra C. Carr [2] and Selene Peng [3]

[1]   Integrated Health Options Ltd., Auckland 1050, New Zealand
[2]   Department of Pathology & Biomedical Science, University of Otago, Christchurch 8140, New Zealand; anitra.carr@otago.ac.nz
[3]   Feedback Research Ltd., Auckland 1050, New Zealand; selene@feedbackresearch.co.nz
*   Correspondence: nicky@integratedhealthoptions.co.nz; Tel.: +64-9524-7745

Received: 6 August 2018; Accepted: 28 August 2018; Published: 30 August 2018

Abstract: Glioblastoma multiforme is a high grade malignant brain tumour with a poor prognosis. Here we report the case of a woman with glioblastoma who lived for over four years from diagnosis (median survival 12 months and 2% survival for three years), experiencing good quality of life for most of that time. She underwent initial debulking craniotomy, radiotherapy and chemotherapy, as well as having intravenous vitamin C infusions 2–3 times weekly over the four years from diagnosis. Her progress was monitored by blood tests, regular computerised tomography (CT) and magnetic resonance imaging (MRI) scans, clinical reviews and European Organization for the Research and Treatment of Cancer quality of life questionnaires (EORTC QLQ C30). Our case report highlights the benefits of intravenous vitamin C as a supportive therapy for patients with glioblastoma.

Keywords: vitamin C; intravenous vitamin C; ascorbic acid; glioblastoma; neoplasms; quality of life

1. Introduction

Glioblastoma multiforme is a high grade malignant brain tumour with a poor prognosis; median survival 12 months and 2% survive for three years [1,2]. We describe the case of a 55-year-old woman who responded much better than expected, with good quality of life for almost four years from diagnosis. She chose to combine conventional treatments with intravenous (IV) vitamin C therapy. Although there are limited clinical trials on the use of IV vitamin C for people with cancer, the available evidence indicates it is safe and generally well tolerated when combined with standard cancer therapies [3–5]. A small trial of IV vitamin C in combination with standard therapy in glioblastoma patients indicated a trend towards enhanced median survival [6]. In our case, the IV vitamin C was well tolerated, quality of life improved markedly in the first year, and then stabilised, and the patient had improved or stable blood tests throughout, including normal renal function.

2. Case Report

The enduring power of attorney provided consent for the presentation of the patient's case. The 55 years old woman was diagnosed in November 2010 with glioblastoma multiforme following a 10 days history of headaches and constipation. Concurrently, she was diagnosed with extranodal marginal zone lymphoma, affecting the left parotid gland and bladder. She was previously fit and healthy, with no significant past medical history, and no history of renal stones. She did not drink alcohol and was a nonsmoker. Her sister died of a brain tumour age 40, and her father of prostate and bladder cancer age 84. She underwent craniotomy and debulking surgery in December 2010, followed by a course of radiotherapy and temozolomide. She had no specific conventional treatment for the lymphoma.

Overall, the patient received 25 radiation treatments March to April 2011, and a further 10 treatments April 2014, and temozolomide for six cycles from March to June 2011, and a further six cycles from April 2014. She had no conventional cancer treatments for almost three years, from July 2011 until April 2014.

2.1. IV Vitamin C Treatment

When the patient first presented to our clinic mid-January 2011 she had fatigue, lethargy and an early chest infection. She was diagnosed with pneumonia the following day, which was treated with IV antibiotics in hospital, at which time she also had a blood transfusion for anaemia. The patient commenced regular IV vitamin C infusions a few days later, approximately three weeks following her brain surgery for glioblastoma multiforme. Baseline blood tests showed normal renal function and normal glucose-6-phosphate dehydrogenase (G6PD) status. If G6PD deficiency is present we do not recommend more than 25 grams IV vitamin C as there is an increased risk of haemolysis. Her non-fasting baseline plasma vitamin C level was 1.2 mg/dL (68 μmol/L); she had consumed a 1 g vitamin C supplement prior to coming to the clinic. The patient received IV vitamin C working up to 85 g/infusion, three times/week for the first six months to June 2011, then twice/week for over three years, stopping October 2014. Post-infusion plasma vitamin C levels were monitored for the first four months after she commenced IV vitamin C infusion to ensure that the proposed ascorbic acid therapeutic level was achieved. The average post-infusion plasma vitamin C level was 393 mg/dL (22 mmol/L), with a dose of 85 g (1.1 g/kg). The patient's renal function tests were normal throughout IV vitamin C treatments with creatinine ranging from 51–67 μmol/L and eGFR mostly >90 mL/min/1.73 m^2. The IV vitamin C was well tolerated.

2.2. Outcome and Follow-Up

CT scans or MRI scans with contrast were carried out every 3–6 months over the four years from diagnosis. These indicated stable disease until March 2014 (Table 1). In July 2011, her extranodal marginal zone lymphoma, eight months after diagnosis with no specific lymphoma treatments, showed stable disease on CT scan of the left parotid gland, with a possible decrease in the size of bladder mass. Six monthly follow-ups reported stable disease until March 2014 when a MRI scan noted lymphoma increase in both orbits, but the comment was made that 'overall lymphoma was reasonably stable'.

Table 1. Computerised tomography (CT) and magnetic resonance imaging (MRI) scans of brain.

Date	CT Scan of Brain	MRI Scan of Brain
July 2011	Residual tumour	
January 2012		Stable
May 2012		Stable
September 2012	Reduced tumour size	
August 2013	Stable	
October 2013		Stable
March 2014		Progression of brain tumour with new and increasing foci
June 2014		Likely disease progression
September 2014		Increase size of brain tumour over 1 year, 4 discrete lesions

Following the initiation of IV vitamin C treatments (January 2011), the patient reported clinical improvements that were maintained for almost four years following diagnosis. She noticed improved energy and walking distance four weeks after initially commencing IV vitamin C, prior to starting her first course of radiation therapy. She chose to continue IV vitamin C throughout her two courses of chemotherapy and radiotherapy, despite being advised that we do not recommend IV vitamin C concurrently with these treatments as there is limited research into benefits or interactions in this area. She felt it helped her symptoms and reported tolerating the radiotherapy 'pretty well'.

The patient's health-related quality of life (QoL) was monitored using the EORTC QLQ C30 for the first 12 months, and subsequently through clinical reviews. Compared to the patient's QoL data prior to commencing IV vitamin C treatment, dramatic improvement was reported in the three months survey following initiation of IV vitamin C treatment (at that time the patient had just completed radiotherapy and started chemotherapy) and the improvement was maintained for the first 12 months (Table 2). Clinical symptoms such as fatigue, dyspnoea, insomnia, appetite loss and diarrhoea resolved and the patient's physical and role functioning increased significantly, as well as 'global health status', which improved from 'very poor' to 'excellent'.

Table 2. Health-related quality of life following intravenous vitamin C (IVC) treatments.

Scales	Before IVC	3 Months after IVC	12 Months after IVC
Symptoms scales			
Fatigue	100	0	0
Nausea	0	0	0
Pain	0	0	0
Dyspnoea	100	0	0
Insomnia	100	0	0
Appetite loss	100	0	0
Constipation	0	0	0
Diarrhoea	67	0	0
Functional scales			
Physical functioning	27	100	100
Role functioning	0	100	100
Emotional functioning	100	100	100
Cognitive functioning	100	100	100
Social Functioning	100	100	100
Global health status	0	100	83

Formal QoL monitoring was not collected after the first 12 months, but from clinical reviews the patient maintained a good QoL, feeling 'very well' until April 2014 when she developed seizures. These were well controlled with medication and she continued working full-time until October 2014. Early November she deteriorated with reduced alertness, right sided weakness, and expressive dysphasia. She ceased IV vitamin C at this time. She continued to deteriorate until 26 February 2015 when she passed away at home, four years and three months from the initial diagnosis.

3. Discussion

The use of pharmacological ascorbate as an adjuvant cancer therapy was proposed as early as 1976 [7]. More recent studies have consistently shown selective toxicity to cancer cells compared to normal cells in vitro and in vivo [6,8–11]. The mechanism of action for selective cancer cell toxicity is not fully understood, although several research studies have suggested that the cytotoxic effect is mediated through ascorbate-mediated reduction of catalytic metal ions and subsequent generation of hydrogen peroxide, which can induce oxidative damage to macromolecules (i.e., DNA, proteins and lipids), deplete cellular adenosine triphosphate (ATP), and thereby cause cell death [12–14]. Another plausible mechanism of action has been recently proposed, as high dose ascorbate administration has been shown to slow tumour growth and decrease hypoxia through suppressing the activity of hypoxia-inducible factor (HIF-1), which is known to contribute to tumour progression [11,15,16]. Clinical trials in the use of IV vitamin C in critically ill patients [17–19] and patients with cancer [3–5,20,21] have demonstrated lack of toxicity, good safety and tolerability [22].

Our case report outlines the unexpected increase in progression free and overall survival in a woman with glioblastoma who had IV vitamin C from shortly after diagnosis, for almost four years, stopping three months before she died. She chose to continue IV vitamin C throughout her two courses of chemotherapy and radiotherapy and her increased survival time supports the findings of a small

Phase 1 clinical trial in the concurrent use of IV vitamin C in patients with glioblastoma receiving chemotherapy and radiotherapy [6]. Although the IV vitamin C was the only additional treatment she was undertaking we cannot conclude that it was the primary factor in her increased progression free survival. However, it is interesting to note that epigenetic dysregulation of the vitamin C-dependent DNA hydroxylase ten-eleven translocation-2 (TET2) is observed in human glioblastoma and decreased hydroxymethylcytosine is associated with shorter malignant glioma survival [23,24], suggesting a possible mechanism by which vitamin C could aid in glioblastoma therapy. Nevertheless, it is likely that the IV vitamin C contributed to the patient's improved quality of life, as has been observed in numerous other studies [25].

4. Conclusions

Our case report indicates that IV vitamin C may be a useful supportive therapy in the management of people with glioblastoma multiforme. IV vitamin C even at doses of over 1 g/kg body weight was very well tolerated. There were no renal stones or impairment of renal function throughout four years of treatment with IV vitamin C in this case. Overall, IV vitamin C given concurrently with chemotherapy and radiotherapy had no apparent adverse reactions and contributed to an improved quality of life and potentially an increase in progression free and overall survival. Further trials on the use of IV vitamin C as an adjuvant therapy in people with glioblastoma are warranted. Currently there is a Phase II trial underway investigating pharmacological vitamin C combined with radiation and temozolomide in glioblastoma multiforme (NCT02344355).

Author Contributions: Conceptualization, N.B.; Investigation, N.B. and S.P.; Writing–Original Draft Preparation, N.B., A.C. and S.P.

Funding: This research received no external funding.

Acknowledgments: A.C. is supported by a Health Research Council of New Zealand Sir Charles Hercus Health Research Fellowship.

Conflicts of Interest: The authors declare no conflict of interest.

References

1. Yersal, O. Clinical outcome of patients with glioblastoma multiforme: Single center experience. *J. Oncol. Sci.* **2017**, *3*, 123–126. [CrossRef]
2. Adeberg, S.; Bostel, T.; König, L.; Welzel, T.; Debus, J.; Combs, S.E. A comparison of long-term survivors and short-term survivors with glioblastoma, subventricular zone involvement: A predictive factor for survival? *Radiat. Oncol.* **2014**, *9*, 1–6. [CrossRef] [PubMed]
3. Monti, D.A.; Mitchell, E.; Bazzan, A.J.; Littman, S.; Zabrecky, G.; Yeo, C.J.; Pillai, M.V.; Newberg, A.B.; Deshmukh, S.; Levine, M. Phase I evaluation of intravenous ascorbic acid in combination with gemcitabine and erlotinib in patients with metastatic pancreatic cancer. *PLoS ONE* **2012**, *7*, e29794. [CrossRef] [PubMed]
4. Welsh, J.L.; Wagner, B.A.; van't Erve, T.J.; Zehr, P.S.; Berg, D.J.; Halfdanarson, T.R.; Yee, N.S.; Bodeker, K.L.; Du, J.; Roberts, L.J., 2nd; et al. Pharmacological ascorbate with gemcitabine for the control of metastatic and node-positive pancreatic cancer (PACMAN): Results from a phase I clinical trial. *Cancer Chemother. Pharmacol.* **2013**, *71*, 765–775. [CrossRef] [PubMed]
5. Hoffer, L.J.; Robitaille, L.; Zakarian, R.; Melnychuk, D.; Kavan, P.; Agulnik, J.; Cohen, V.; Small, D.; Miller, W.H., Jr. High-dose intravenous vitamin C combined with cytotoxic chemotherapy in patients with advanced cancer: A Phase I-II clinical trial. *PLoS ONE* **2015**, *10*, e0120228. [CrossRef] [PubMed]
6. Schoenfeld, J.D.; Sibenaller, Z.A.; Mapuskar, K.A.; Wagner, B.A.; Cramer-Morales, K.L.; Furqan, M.; Sandhu, S.; Carlisle, T.L.; Smith, M.C.; Abu Hejleh, T.; et al. O_2- and H_2O_2-Mediated Disruption of Fe Metabolism Causes the Differential Susceptibility of NSCLC and GBM Cancer Cells to Pharmacological Ascorbate. *Cancer Cell* **2017**, *31*, 487–500.e8. [CrossRef] [PubMed]
7. Cameron, E.; Pauling, L. Supplemental ascorbate in the supportive treatment of cancer: Prolongation of survival times in terminal human cancer. *Proc. Natl. Acad. Sci. USA* **1976**, *73*, 3685–3689. [CrossRef] [PubMed]

8. Du, J.; Martin, S.M.; Levine, M.; Wagner, B.A.; Buettner, G.R.; Wang, S.H.; Taghiyev, A.F.; Du, C.; Knudson, C.M.; Cullen, J.J. Mechanisms of ascorbate-induced cytotoxicity in pancreatic cancer. *Clin. Cancer Res.* **2010**, *16*, 509–520. [CrossRef] [PubMed]
9. Ma, Y.; Chapman, J.; Levine, M.; Polireddy, K.; Drisko, J.; Chen, Q. High-dose parenteral ascorbate enhanced chemosensitivity of ovarian cancer and reduced toxicity of chemotherapy. *Sci. Transl. Med.* **2014**, *6*, 222ra18. [CrossRef] [PubMed]
10. Yun, J.; Mullarky, E.; Lu, C.; Bosch, K.N.; Kavalier, A.; Rivera, K.; Roper, J.; Chio, I.I.; Giannopoulou, E.G.; Rago, C.; et al. Vitamin C selectively kills KRAS and BRAF mutant colorectal cancer cells by targeting GAPDH. *Science* **2015**, *350*, 1391–1396. [CrossRef] [PubMed]
11. Campbell, E.J.; Vissers, M.C.; Wohlrab, C.; Hicks, K.O.; Strother, R.M.; Bozonet, S.M.; Robinson, B.A.; Dachs, G.U. Pharmacokinetic and anti-cancer properties of high dose ascorbate in solid tumours of ascorbate-dependent mice. *Free Radic. Biol. Med.* **2016**, *99*, 451–462. [CrossRef] [PubMed]
12. Chen, Q.; Espey, M.G.; Krishna, M.C.; Mitchell, J.B.; Corpe, C.P.; Buettner, G.R.; Shacter, E.; Levine, M. Pharmacologic ascorbic acid concentrations selectively kill cancer cells: Action as a pro-drug to deliver hydrogen peroxide to tissues. *Proc. Natl. Acad. Sci. USA* **2005**, *102*, 13604–13609. [CrossRef] [PubMed]
13. Chen, Q.; Espey, M.G.; Sun, A.Y.; Lee, J.H.; Krishna, M.C.; Shacter, E.; Choyke, P.L.; Pooput, C.; Kirk, K.L.; Buettner, G.R.; et al. Ascorbate in pharmacologic concentrations selectively generates ascorbate radical and hydrogen peroxide in extracellular fluid in vivo. *Proc. Natl. Acad. Sci. USA* **2007**, *104*, 8749–8754. [CrossRef] [PubMed]
14. Olney, K.E.; Du, J.; van't Erve, T.J.; Witmer, J.R.; Sibenaller, Z.A.; Wagner, B.A.; Buettner, G.R.; Cullen, J.J. Inhibitors of hydroperoxide metabolism enhance ascorbate-induced cytotoxicity. *Free Radic. Res.* **2013**, *47*, 154–163. [CrossRef] [PubMed]
15. Fischer, A.P.; Miles, S.L. Ascorbic acid, but not dehydroascorbic acid increases intracellular vitamin C content to decrease Hypoxia Inducible Factor -1 alpha activity and reduce malignant potential in human melanoma. *Biomed. Pharmacother.* **2017**, *86*, 502–513. [CrossRef] [PubMed]
16. Wilkes, J.G.; O'Leary, B.R.; Du, J.; Klinger, A.R.; Sibenaller, Z.A.; Doskey, C.M.; Gibson-Corley, K.N.; Alexander, M.S.; Tsai, S.; Buettner, G.R.; et al. Pharmacologic ascorbate (P-AscH(-)) suppresses hypoxia-inducible Factor-1alpha (HIF-1alpha) in pancreatic adenocarcinoma. *Clin. Exp. Metastasis* **2018**. [CrossRef] [PubMed]
17. Fowler, A.A.; Syed, A.A.; Knowlson, S.; Sculthorpe, R.; Farthing, D.; DeWilde, C.; Farthing, C.A.; Larus, T.L.; Martin, E.; Brophy, D.F.; et al. Phase I safety trial of intravenous ascorbic acid in patients with severe sepsis. *J. Transl. Med.* **2014**, *12*, 32. [CrossRef] [PubMed]
18. Zabet, M.H.; Mohammadi, M.; Ramezani, M.; Khalili, H. Effect of high-dose Ascorbic acid on vasopressor's requirement in septic shock. *J. Res. Pharm. Pract.* **2016**, *5*, 94–100. [PubMed]
19. Tanaka, H.; Matsuda, T.; Miyagantani, Y.; Yukioka, T.; Matsuda, H.; Shimazaki, S. Reduction of resuscitation fluid volumes in severely burned patients using ascorbic acid administration: A randomized, prospective study. *Arch. Surg.* **2000**, *135*, 326–331. [CrossRef] [PubMed]
20. Polireddy, K.; Dong, R.; Reed, G.; Yu, J.; Chen, P.; Williamson, S.; Violet, P.C.; Pessetto, Z.; Godwin, A.K.; Fan, F.; et al. High dose parenteral ascorbate inhibited pancreatic cancer growth and metastasis: Mechanisms and a Phase I/IIa study. *Sci. Rep.* **2017**, *7*, 17188. [CrossRef] [PubMed]
21. Stephenson, C.M.; Levin, R.D.; Spector, T.; Lis, C.G. Phase I clinical trial to evaluate the safety, tolerability, and pharmacokinetics of high-dose intravenous ascorbic acid in patients with advanced cancer. *Cancer Chemother. Pharmacol.* **2013**, *72*, 139–146. [CrossRef] [PubMed]
22. Nauman, G.; Gray, J.C.; Parkinson, R.; Levine, M.; Paller, C.J. Systematic review of intravenous ascorbate in cancer clinical trials. *Antioxidants* **2018**, *7*, 89. [CrossRef] [PubMed]
23. Garcia, M.G.; Carella, A.; Urdinguio, R.G.; Bayon, G.F.; Lopez, V.; Tejedor, J.R.; Sierra, M.I.; Garcia-Torano, E.; Santamarina, P.; Perez, R.F.; et al. Epigenetic dysregulation of TET2 in human glioblastoma. *Oncotarget* **2018**, *9*, 25922–25934. [CrossRef] [PubMed]

24. Orr, B.A.; Haffner, M.C.; Nelson, W.G.; Yegnasubramanian, S.; Eberhart, C.G. Decreased 5-hydroxymethylcytosine is associated with neural progenitor phenotype in normal brain and shorter survival in malignant glioma. *PLoS ONE* **2012**, *7*, e41036. [CrossRef] [PubMed]
25. Carr, A.C.; Vissers, M.C.M.; Cook, J.S. The effect of intravenous vitamin C on cancer- and chemotherapy-related fatigue and quality of life. *Front. Oncol.* **2014**, *4*, 1–7. [CrossRef] [PubMed]

![antioxidants logo] *antioxidants*

MDPI

Article

High Dose Ascorbate Causes Both Genotoxic and Metabolic Stress in Glioma Cells

Maria Leticia Castro [1], Georgia M. Carson [1], Melanie J. McConnell [1,2] and Patries M. Herst [2,3,*]

[1] School of Biological Sciences, Victoria University, P.O.Box 600, Wellington 6140, New Zealand; leticia.castro@vuw.ac.nz (M.L.C.); georgia.carson@hotmail.co.nz (G.M.C.); Melanie.McConnell@vuw.ac.nz (M.J.M.)
[2] Malaghan Institute of Medical Research, P.O.Box 7060, Wellington 6242, New Zealand
[3] Department of Radiation Therapy, University of Otago, P.O.Box 7343, Wellington 6242, New Zealand
* Correspondence: patries.herst@otago.ac.nz; Tel.: +64-4-3855475; Fax: +64-4-43855375

Received: 10 May 2017; Accepted: 19 July 2017; Published: 22 July 2017

Abstract: We have previously shown that exposure to high dose ascorbate causes double stranded breaks (DSBs) and a build-up in S-phase in glioblastoma (GBM) cell lines. Here we investigated whether or not this was due to genotoxic stress as well as metabolic stress generated by exposure to high dose ascorbate, radiation, ascorbate plus radiation and H_2O_2 in established and primary GBM cell lines. Genotoxic stress was measured as phosphorylation of the variant histone protein, H2AX, 8-oxo-7,8-dihydroguanine (8OH-dG) positive cells and cells with comet tails. Metabolic stress was measured as a decrease in NADH flux, mitochondrial membrane potential (by CMXRos), ATP levels (by ATP luminescence) and mitochondrial superoxide production (by mitoSOX). High dose ascorbate, ascorbate plus radiation, and H_2O_2 treatments induced both genotoxic and metabolic stress. Exposure to high dose ascorbate blocked DNA synthesis in both DNA damaged and undamaged cell of ascorbate sensitive GBM cell lines. H_2O_2 treatment blocked DNA synthesis in all cell lines with and without DNA damage. DNA synthesis arrest in cells with damaged DNA is likely due to both genotoxic and metabolic stress. However, arrest in DNA synthesis in cells with undamaged DNA is likely due to oxidative damage to components of the mitochondrial energy metabolism pathway.

Keywords: high dose ascorbate; H_2O_2; radiation; oxidative stress; genotoxic stress; metabolic stress; DNA synthesis arrest

---

1. Introduction

The last decade has seen a renewed interest in intravenous high dose (pharmacological) ascorbate (AA) as an anticancer treatment. Most authors in the field attribute the anticancer effect of high dose AA to its pro-oxidant effect. In the extracellular acidic and metal-rich tumour environment, high dose AA generates extracellular hydrogen peroxide (H_2O_2), which diffuses into adjacent cancer cells and overwhelms the anti-oxidant defence system (reviewed by [1]). The resulting oxidative stress damages macromolecules [1] as well as depleting NAD^+ and ATP levels [2–5]. Yun and colleagues reported recently that the oxidised form of AA, dehydroascorbate (DHA) rather than AA was responsible for selectively killing glycolysis-driven colorectal cancer cells with BRAF and KRAS mutations [6]. DHA is transported into cells through glucose transporters (GLUT-1), where it is reduced back to AA at the expense of glutathione (GSH), causing oxidative stress and inhibition of glyceraldehyde-3-phosphate dehydrogenase (GAPDH) and thus glycolysis, leading to ATP depletion [6,7]. In addition to causing oxidative stress, AA has been shown to increase hypoxia inducible factor, HIF-1, hydroxylase activity, leading to a decrease in HIF-1 pathway activation and a less aggressive phenotype in colorectal [8] and endometrial cancer [9], and inhibit the proliferation of breast cancer MCF-7 mammospheres [10].

Most authors have reported that high dose AA has little or no effect on non-cancerous cell lines (reviewed by [11]) and few side effects in animal models [12,13] or clinical trials [4,14,15]. Cancer specificity has been attributed to the acidic and metal-rich tumour micro-environment combined with the inferior anti-oxidant capacity and compromised DNA repair pathways of tumour cells [16–19]. Combining high dose AA with ionizing radiation should therefore radio-sensitize highly radiation resistant glioblastoma (GBM) cells [20,21] and improve the dismal prognosis for GBM patients [22].

Our group has studied the effect of high dose AA on radio-sensitization of GBMs in several studies. We initially showed that a GBM cell line isolated from a GBM patient was much more sensitive to high dose AA, radiation and combined treatments than a mouse-derived normal glial cell line [23]. However, a subsequent more detailed study showed that six human GBM cell lines, a human glial cell line and human umbilical vein endothelial cells were similarly sensitive to high dose AA and/or radiation. Sensitivity depended on their antioxidant and DNA repair capacity regardless of their cancerous status [24]. We further showed that exposure to high dose AA caused accumulation in S-phase as well as genotoxic stress. Genotoxic stress was demonstrated by a higher percentage of cells with foci caused by phosphorylation of the variant histone protein (H2AX) associated with the DNA damage response (γH2AX) as well as more γH2AX foci per cell [23,24]. The number of γH2AX foci correlates well with the number of double strand breaks (DSBs) generated by ionising radiation [25–28]. However, fewer than half of γH2AX foci induced by H_2O_2 [25–27] and UV [28] are associated with DSBs; with foci produced during replication likely representing stalled replication forks, which can either be repaired or progress to DSBs [25–28]. Another type of DNA lesion that causes genotoxic stress are 8-oxo-7,8-dihydroguanine (8OH-dG) lesions caused by aggressive hydroxyl free radicals generated by H_2O_2 in the close vicinity of DNA [18]. Although these lesions are also present in some untreated GBM cell lines and many GBM tumours [21], they are generated specifically in response to H_2O_2 [25–27]. The lesions are rapidly repaired by base excision repair (BER) and if unrepaired may generate single stranded breaks [18] or double stranded breaks [25–28]. Metabolic stress, in the form of low NAD^+ and low ATP levels as a result of high dose AA exposure, has been shown by several authors [2–5]. In this paper, we analysed the extent of genotoxic and metabolic stress and the effect on DNA synthesis by high dose AA in established and patient-derived GBM cell lines and compared these effects to those of radiation, H_2O_2 and combined treatments.

2. Materials and Methods

2.1. Materials

Unless otherwise noted, tissue plasticware was purchased from Corning (In Vitro Technologies, Auckland, New Zealand); all cell culture reagents were from Gibco BRL (Thermo Fisher Scientific, Auckland, New Zealand). Alexa Fluor 488 anti-H2AX-Phosphorylated (Ser139) Antibody was from BioLegend (Norrie Biotech, Auckland, New Zealand). Rabbit anti-8-OHdG polyclonal antibody (J-1: sc-139586) was from, Santa Cruz Biotech (Dallas, Texas, USA) and isotype control (IgG/10500C) was from Thermofisher Scientific (Wellington, New Zealand). Goat Polyclonal Anti-Rabbit IgG H&L (Alexa Fluor® 488) secondary antibody was from Abcam (Cambridge, MA, USA). Foxp3/Transcription Factor Fixation/Permeabilization Concentrate and Diluent was purchased from eBioscience (Huntingtree Bioscience Supplies, Auckland, New Zealand). Click-iT EdU Alexa Fluor Flow Cytometry Assay Kits, Vybrant DyeCycle Stains, MitoSOX™ Red Mitochondrial Superoxide Indicator and MitoTracker® Red CMXRos were purchased from Life Technologies (Thermo Fisher Scientific, Auckland, New Zealand). CellTiter 96® AQueous One Solution Cell Proliferation Assay (MTS) was sourced from Promega Corporation (Madison, WI, USA). Luminescent ATP Detection Assay Kit was obtained from Abcam (Cambridge, MA, USA). Sodium ascorbate, and all other chemicals and reagents were from Sigma Chemical Company (St. Louis, MO, USA).

2.2. Cell Lines

GBM cell lines, LN18, U87MG and T98G were obtained from the American Type Culture Collection. Primary GBM cells (NZG0809, NZG1003) were isolated and cultured from GBM material as previously described [29]. GBMs were grown in RPMI-1640 supplemented with 5% (*v*/*v*) FBS. All cells were maintained in a humidified incubator at 37 °C/5% CO_2.

2.3. Ascorbate Treatment

Exponentially growing cells (30–40% confluent) were seeded 24 h prior to treatment in 6 well plates (3–5 × 10^4 cells/well). Cells were exposed to 5 mM AA in media for 1 h, washed in Dulbecco's Phosphate Buffer Saline (PBS, 1.4 M NaCl, 27 mM KCl, 170 mM NaH_2PO_4, 17.6 mM KH_2PO_4) and re-incubated in fresh medium. Cells that received radiation were irradiated in the presence of AA.

2.4. Radiation Treatment

Exponentially growing cells (30–40% confluent) were irradiated fully immersed in medium with 6 Gy using Cesium-137 γ-rays (Gammacell 3000 Elan, Best Theratronics, Kanata, ON, Canada). After irradiation, cells were re-incubated in fresh medium.

2.5. DNA Synthesis: EdU Incorporation

Fluorescent detection of incorporated thymidine analogue EdU (5-ethynyl-2'-deoxyuridine) was used as a measure of DNA synthesis. Cells were pulse-labelled with 4 μM EdU for 1 h prior to analysis. Incorporated EdU was detected using a copper catalyzed covalent reaction between Click-iT® EdU Alexa Fluor® 488 or Alexa Fluor® 647 dye azide and an alkyne on the ethynyl moiety of EdU. Briefly, cells were pulse labelled with EdU for 1hr prior to harvesting, washed twice in FACS buffer (PBS + 1% Bovine Serum Albumin (BSA), and incubated in Click-iT fixative at room temperature for 15 min in the dark. Cells were washed and resuspended in saponin-based permeabilisation buffer for 15 min prior to incubation in the Click-iT AF647 reaction cocktail for 30 min in the dark at room temperature. For DNA content staining, these cells were incubated in a 5 μM Vybrant Dye Cycle Green/HBSS staining solution for 30 min at 37 °C prior to analysis by flow cytometry using a BD FACSCanto II (Becton Dickinson, San Jose, CA, USA). Data were analysed using FlowJo (TreeStar, Ashland, OR, USA).

2.6. γH2AX Labelling

Genotoxic stress was determined by measuring the extent of phosphorylation of the histone protein, H2AX, using Alexa Fluor 488 labelled anti-γH2AX antibody in permeabilised cells. EdU labelled cells were harvested and washed in PBS buffer, distributed into 96 well plates (5 × 10^5 cells/well), washed in 200 μL FACS buffer (PBS + 1% BSA), pelleted at 400× *g* (in a Megafuge 2.0R, Heraeus centrifuge, Labcare, Buckinghamshire, UK) for 4 min and fixed in 200 μL Forkhead box P3 (FoxP3)/Transcription Factor Fixation/ Permeabilization solution for 45 min at room temperature. Cells were washed twice in 200 μL 1x Permeabilization/Wash buffer (PBS, 1% BSA, 0.5% Saponin), pelleted and incubated in Click-iT® AF647 reaction cocktail for 30 min in the dark. Cells were then washed twice in Perm/Wash buffer and incubated in antibody solution (50 μL anti-γH2AX antibody or isotype control diluted 1:200 in 1x Perm/Wash buffer) at 4 °C overnight. Following this, cells were washed twice in Perm/Wash buffer and resuspended in 400 μL FACS buffer. Cells were analysed by flow cytometry using a BD FACSCanto II (Becton Dickinson, San Jose, CA, USA) and FlowJo software (TreeStar, Ashland, OR, USA).

2.7. 8OH-dG Lesions

Harvested cells were washed in PBS buffer and distributed into 96 well plates (5 × 10^5 cells/well), washed in 200 μL FACS buffer (PBS + 1% BSA), pelleted at 400× *g* (in a Megafuge 2.0R, Heraeus centrifuge, Labcare, Buckinghamshire, UK) for 4 min and fixed in 200 μL Foxp3/Transcription Factor Fix/Perm solution for 45 min at room temperature. Cells were washed twice in 200 μL 1x Permeabilization/Wash buffer (PBS, 1% BSA, 0.5% Saponin), pelleted and resuspended in primary antibody solution (50 μL rabbit anti-8OH-dG polyclonal antibody or isotype control (diluted at 400 ng/mL in 1x Perm/Wash buffer) at 4 °C overnight, washed twice in Perm/Wash buffer and resuspended in secondary antibody solution (Goat Polyclonal Anti-Rabbit IgG H&L (Alexa Fluor® 488), Abcam) at 1000 ng/mL) for an hour at room temperature. Cells were then washed twice in Perm/wash buffer and resuspended in 400 μL FACS buffer for analysis by flow cytometry using a BD FACSCanto II (Becton Dickinson, San Jose, CA, USA) and FlowJo software (TreeStar, Ashland, OR, USA).

2.8. Comet Tail Assay

Glass slides (LabServ Superfrost Plus) were pre-coated with 1% normal melting point agarose (Invitrogen UltraPure Agarose) and air-dried for 24 h. Cells were added to 1% low melting point agarose at a ratio of 1:10 (v/v) to final cell concentration 1×10^4 cells/mL, and dropped onto agarose coated slides. Agarose and cells were air dried for 30 min at room temperature, then lysed in pre-chilled lysis solution (2.5 M NaCl, 100 mM EDTA pH 10, 10 mM Trizma, 1% sodium lauryl sarcosinate, and 1% Triton X-100, pH 10) for 1 h at 4 °C. Slides were rinsed in 1x Tris-Borate-EDTA buffer TBE, and equilibrated for 30 min in 1x TBE before electrophoresis at 30 volts/cm for 60 min. Slides were stained with 10 μg/mL propidium iodide for 20 min at 4 °C, rinsed in TBE, and imaged on a fluorescent microscope (Olympus BX51 microscope with TXRED filter). Cell nuclei were analysed using ImageJ Comet Assay plugin, based on an NIH Image Comet Assay by H.M. Miller (https://www.med.unc.edu/microscopy/resources/imagej-plugins-and-macros/comet-assay) from Robert Bagnell. Briefly, tight ovals are drawn around the head and the tail. The tail length is the distance from the centre of the head (defined as the average of XY coordinates of all pixels in the head oval) to the centre of mass of the tail (defined as the brightness-weighted average of XY coordinates of the selected tail oval). Tail length is calculated as the Pythagorean distance between the two points. ImageJ outputs were imported into Microsoft Excel and averages were imported into Prism (Graphpad) V6.0 for analysis.

2.9. MTS Assay

The colorimetric CellTiter 96® AQueous One Solution Cell Proliferation Assay containing the soluble tetrazolium compound MTS (3-(4,5-dimethylthiazol-2-yl)-5-(3-carboxymethoxyphenyl) -2-(4-sulfophenyl)-2H-tetrazolium, inner salt) and electron coupling reagent PES (phenazine ethosulfate) were utilised for assessment of NADH flux in cells. Cells (5×10^3 cells/well) were plated into 96 well plates and incubated overnight. Cells were treated and washed three times in 300 μL of PBS. 100 μL of culture media was replaced and 20 μL of CellTiter 96® AQueous One Solution Reagent was added. Plates were incubated for 1−4 h at 37 °C and absorbance was measured at 490 nm on an Enspire 2300 plate reader (Perkin-Elmer, Shelton, CT, USA).

2.10. MitoTracker® Red CMXRos

Mitochondrial membrane potential was assessed using the cell-permeant X-rosamine derived MitoTracker® Red CMXRos dye. Treated cells and untreated cells were harvested and incubated in 50 nM CMXRos in PBS for 30 min at 37 °C. Cells were analysed by flow cytometry using a BD FACSCanto II (Becton Dickinson, San Jose, CA, USA) and FlowJo software (TreeStar, OR, USA).

2.11. MitoSOX™ Red Mitochondrial Superoxide Indicator

Live-cell permeant MitoSOX Red superoxide indicator targets mitochondria, where it is selectively oxidized by mitochondrial superoxide to exhibit red fluorescence. Treated and untreated cells were harvested and incubated in 5 μM MitoSOX in PBS for 10 min at 37 °C, fluorescent signal detected by flow cytometry using a FACSCanto II (Becton Dickinson, San Jose, CA, USA) with FlowJo software (TreeStar, OR, USA).

2.12. ATP Measurements

Cellular ATP was assessed using the firefly luciferase based Luminescent ATP Detection Assay Kit (Abcam, Cambridge, UK) according to manufacturer's instruction. Cells (2×10^3 cells/well) were plated into 96 well cell culture treated white plates and incubated overnight. Cells were treated, washed, and lysed in 50 μL of detergent for 5 min on an orbital shaker at 700 rpm. Substrate solution (50 μL) was added and cells were incubated in the dark for a further 5 min on an orbital shaker at 700 rpm. The plate was dark adapted for 10 min prior to luminescent reading using an Enspire plate reader (Perkin-Elmer, Shelton, CT, USA).

2.13. Cell Viability

GL261 cells were collected 48 h after treatment by trypsinization, washed in PBS and resuspended in 1 μg/mL propidium iodide, for cell count and dye exclusion using flow cytometry using a BD FACSCanto II (Becton Dickinson, San Jose, CA, USA) and FlowJo software (TreeStar, Ashland, OR, USA). All viability assays were completed at least three times in triplicate.

2.14. Statistical Analysis

Data were analysed using Excel (Microsoft v. 2010; Redmond Campus, Redmond, WA, USA) or Prism (Graphpad) V6.0. All experiments were done at least three times in triplicate; values are averages ± standard error of the means (SEM). Flow cytometry plots are representative of at least 3 separate experiments.

3. Results

3.1. GBM Cells Have Different Sensitivities to High Dose Ascorbate

Our previous research showed that TG98G cells were less sensitive to 5 mM AA as evidenced by higher clonogenicity than other GBM cells, most likely because of its very high antioxidant content [26]. A more in-depth analysis of AA sensitivity (Figure 1) showed that the IC_{50} of T98G cells was much higher (16.5 mM AA) than that of LN18, NZG0809 and NZG1003 cells (5.8, 6.8 and 7.6 mM AA respectively).

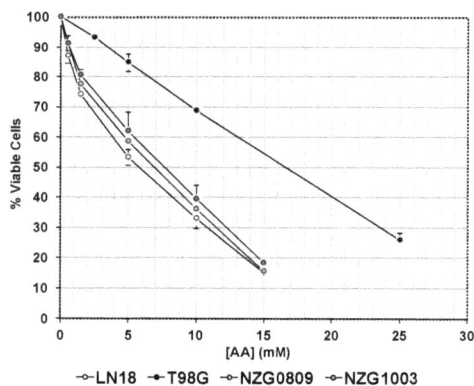

Figure 1. Sensitivities of different glioblastoma (GBM) cells to high dose ascorbic acid (AA). Cells were exposed to different concentrations of AA. Viability was measured by Trypan blue exclusion after 48 h. Values are averages ± standard error of the means (SEM) of at least 3 independent experiments in triplicate.

3.2. High Dose Ascorbate Generates Oxidative Damage and Double-Stranded DNA Breaks

We have previously used γH2AX foci as a measure of DSBs [23,24]. However, less than half of γH2AX foci after H_2O_2 exposure correlate closely with DSBs [25–27]. As most of the high dose AA effects are thought to be mediated though H_2O_2 formation and H_2O_2 has been shown to generate 8OH-dG lesions, we determined the level of 8OH-dG lesions in our cell lines before and after exposure to 5 mM AA, 6 Gy radiation, a combination of AA and radiation and 500 μM H_2O_2. We saw very few lesions 1–2 h after treatments (results not shown) but increasing numbers after 48 h (Figure 2A,B). We performed single cell gel electrophoresis to confirm that high dose AA does generate some DSBs, as evidenced by the presence of "tails" in the comet tail assay (Figure 2C,D). Notably, there were cells without comet tails (and thus DSBs) after exposure to AA, consistent with previous data demonstrating the presence of γH2AX-negative cells [24].

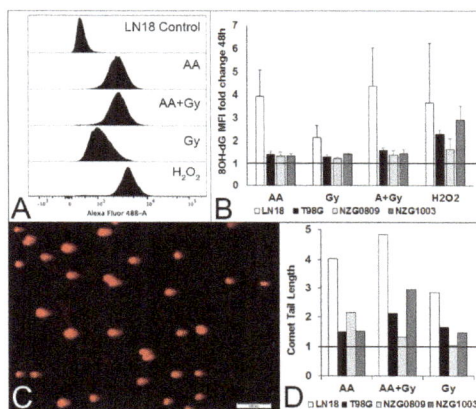

Figure 2. Different types of DNA damage after a 1h exposure to 5 mM AA, 6 Gy, 5 mM AA + 6 Gy and 500 μM H_2O_2. (**A**) Histograms of 8OH-dG lesions 48 h after treatments of LN18 cells. (**B**) Median fluorescence intensity (MFI) fold change compared with untreated cells of LN18, T98G, NZG0809 and NZG1003 cells. Values are averages ± SEM of at least 3 independent experiments in triplicate. Increased 8OH-dG lesions for H_2O_2 treatment over AA, Gy and AA + Gy was statistically significant ($p < 0.05$: unpaired two-tailed student *t*-test) for T98G and NZG1003. C. Representative photograph of comet tails (DSBs) of LN18 cells after a 1 h exposure to 5 mM AA. D. Fold change compared with controls of comet tail length of the different cell lines after a 1 h exposure to 5 mM AA, 6 Gy, 5 mM AA + 6 Gy compared to untreated controls (set at 1). Average number of comet tails measured per cell line: 278 (controls), 162 (AA), 220 (Gy) and 272 (AA + Gy).

3.3. High Dose Ascorbate Abrogates DNA Replication Which Does Not Resolve over Time

We previously showed that GBM cells accumulate in S-phase 24 h after transient exposure to high dose AA [23,24], suggestive of replication fork collapse [25–28]. Here, we measured the extent of DNA synthesis by incorporation of the modified nucleotide ethynyl deoxyuridine (EdU) at early (Figure 3A) and late (Figure 3B) time points after treatments. Plotting EdU incorporation against DNA content resulted in a typical distribution for dividing cells, where the strongly EdU positive cells were predominantly in S-phase with an intermediate DNA content. Strikingly, a 1 h AA exposure completely arrested DNA replication within 2 h in three of the four cell types tested, which was not resolved by 96 h. DNA synthesis in T98G was much less affected by AA, consistent with previous data suggesting this cell line is less sensitive to AA (Figure 1, [24]. In most cells, radiation resulted in a G2 arrest by 24 h with a resumption of normal cell cycle by later time points, in stark contrast to the sustained AA-induced block in DNA synthesis. NZG0809 showed low level of EdU incorporation after AA exposure which was not specifically associated with S-phase.

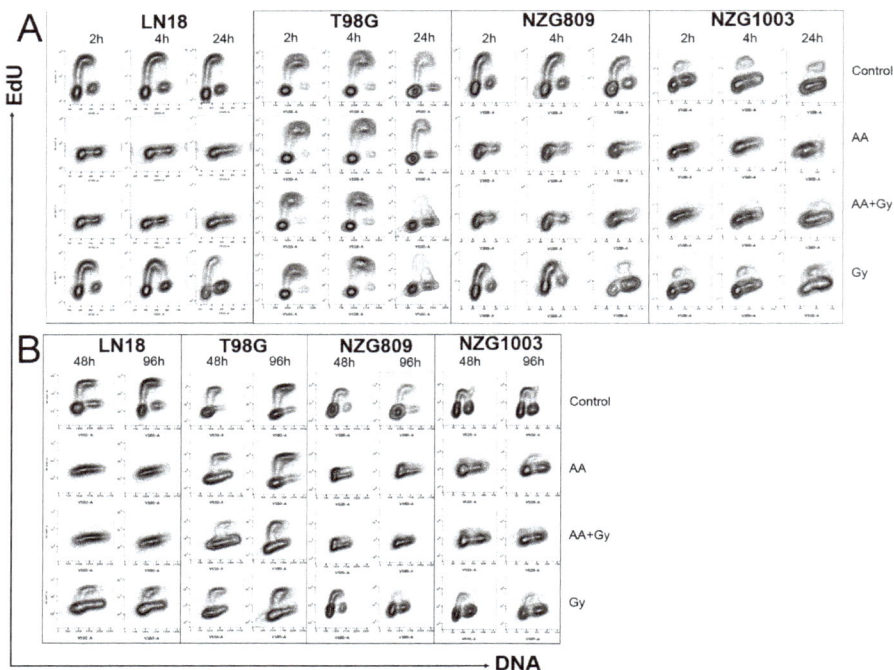

Figure 3. DNA synthesis indicated by EdU incorporation (*y*-axis) and DNA content by nucleic acid content (*x*-axis). G1 cells, with half the content of G2 cells are EdU negative, whereas cells undergoing synthesis have incorporated EdU into the newly synthesised DNA with subsequent increased signal. Cells were harvested at early time points (**A**) and late time points (**B**) after treatments. Flow cytometry plots are representative of at least three independent experiments.

3.4. High Dose Ascorbate Blocks Replication in Both Damaged and Undamaged Cells

The loss of DNA synthesis in AA-treated cells was profound and long-lasting with no recovery of affected cell lines over a four day period. However, the DNA damage data indicated that a small proportion of cells did not sustain any DNA damage, neither γH2AX, DSBs nor 8OH-dG. We therefore directly compared DNA damage with DNA synthesis. DNA damage in untreated controls was visible in both replicating and non-replicating cells, as expected from genetically unstable GBMs [21]. Both damaged and undamaged cells continued to synthesize DNA after radiation. However, almost no DNA synthesis was seen after AA or combined treatment, not even in undamaged cells in LN18 and NZG0809. A small amount of DNA synthesis was seen in NZG1003, particularly at later time points in DNA damaged cells. Despite sustaining moderate amounts of DNA damage, DNA synthesis in T98G was relatively unaffected by AA treatment (Figure 4A). Interestingly, a 1 h exposure to 50–100 μM H_2O_2 stopped DNA synthesis, even in T98G (Figure 4B).

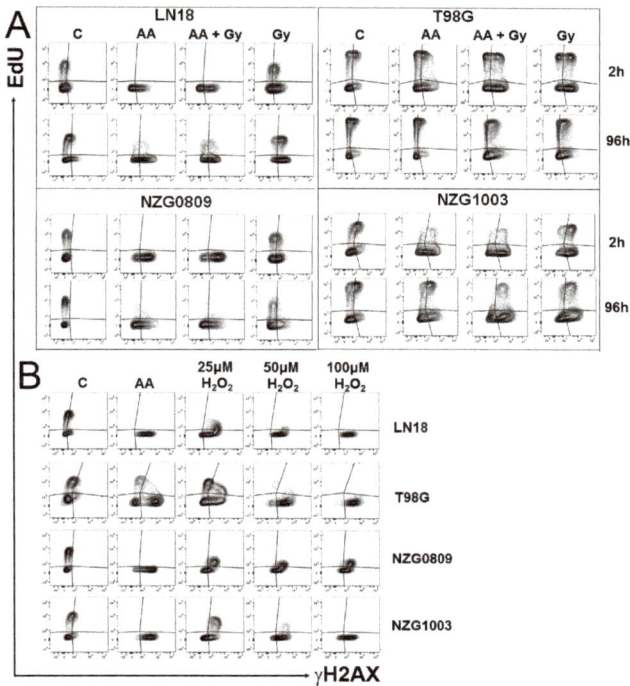

Figure 4. DNA synthesis indicated by EdU incorporation (*y*-axis) versus DNA damage by γH2AX foci (*x*-axis). Cells were treated as indicated and harvested after 2 h (A and B) or 96 h (**A**). (**B**) Effect of a 1 h exposure to 5 mM AA compared with that of different concentrations of H_2O_2. Flow cytometry plots are representative of at least three independent experiments.

3.5. High Dose Ascorbate Causes Metabolic Stress

Cells with damaged DNA can no longer replicate because of genotoxic stress. However, the observation that replication had also stopped in cells without apparent genotoxic stress was unexpected. It suggests that DNA damage, at least for some cells, is not the primary driver to replication loss, and subsequent cell death. High dose AA has also been shown to decrease ATP levels [3–6,16]. We also found that ATP levels declined significantly after a 1 h exposure to 5 mM AA in LN18, NZG0809 and NZG1003 but not in T98G. In comparison, 500 µM H_2O_2 decreased ATP levels more than 5 mM AA in all cell lines, including T98G (Figure 5A). We next determined the effect of treatments on cellular NADH flux by measuring reduction of the water soluble tetrazolium salt, MTS [30]. We found a substantial decrease in MTS reduction 1 h after AA treatment, combined treatment and after 500 µM H_2O_2 in LN18, NZG0809 and NZG1003. MTS reduction in T98G was only minimally affected by 5 mM AA but strongly inhibited by 500 µM H_2O_2 (Figure 5B). Robust mitochondrial electron transport (MET) activity results in a strong potential across the inner mitochondrial membrane. We found a decrease in mitochondrial membrane potential in LN18, NZG0809 and NZG1003 1 h after exposure to AA treatments, whereas H_2O_2 treatment decreased the mitochondrial membrane potential in all four cell lines (Figure 5C). Interestingly, we found an increase in mitochondrial superoxide levels after exposure to AA and H_2O_2 treatments (Figure 5D). Exposure to radiation did not affect ATP levels, MTS reduction, mitochondrial membrane potential or mitochondrial superoxide production in any of the cell lines at such early time points. This was not unexpected as the effects of radiation only become evident at later time points [20]. We previously measured viability 48 h after treatments and found that viability of LN18, NZG0809 and NZG1003 cells decreased substantially after AA treatments. T98G was relatively insensitive to AA whereas NZG1003 was relatively insensitive to radiation (Figure 5E). Correlational analysis (Figure 5F) showed that 1 h after exposure to AA, cellular ATP levels strongly correlated with MTS reduction and mitochondrial membrane potential; MTS reduction strongly correlated with mitochondrial membrane potential; 1 h ATP levels and MTS reduction correlated strongly with 48 h survival.

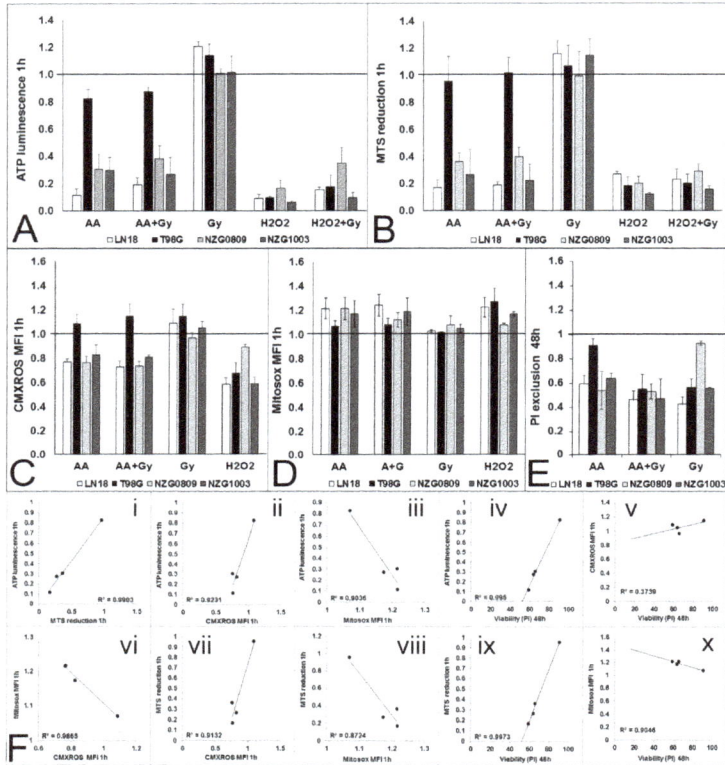

Figure 5. Metabolic stress after 1 h exposure to 5 mM AA, 6 Gy, 5 mM AA + 6 Gy and 500 μM H_2O_2 compared to untreated controls (set at 1). (**A**) Cellular ATP levels measured by luminescence; (**B**) Cellular NADH flux measured by MTS reduction; (**C**) Mitochondrial membrane potential measured by CMXROS; (**D**) Mitochondrial superoxide production measured by Mitosox. (**E**) Viability measured as PI exclusion 48 h after treatments. Fold change compared with untreated control cells. ATP luminescence, MTS reduction and CMXRos MFI were all significantly higher after 6 Gy treatment compared with all other treatments for LN18, NZG0809 and NZG1003. Viability after 5 mM AA treatment was significantly higher for T98G and after 6 Gy treatment for NZG0809. ($p < 0.05$: un-paired two-tailed student t-test). Values are averages ± SEM of at least 3 independent experiments in triplicate. (**F**) Correlations between cellular ATP levels, MTS reduction, mitochondrial membrane potential, mitochondrial superoxide production (all at 1 h) and cell viability at 48 h. Each dot represents one of the four cell lines.

4. Discussion

We previously showed that high dose AA generates γH2AX lesions and causes accumulation of cells in S phase [23,24]. In this paper, we analysed the extent of genotoxic and metabolic stress and the effect on DNA synthesis by high dose AA, radiation and combined treatments. With respect to genotoxic stress, we verified that high dose AA can indeed generate DSBs (as measured by comet tail assay) as well as 8OH-dG lesions. The increase in 8OH-dG lesions over time suggests effective base excision repair at early time points, but sustained free radical production combined with a lack of effective base excision repair at later time points. This may be caused by low cellular ATP levels. The small amount of EdU incorporation in NZG0809 throughout the cell cycle after AA exposure most likely reflects DNA synthesis as part of base excision repair, or other repair mechanisms, rather than DNA replication [31].

DNA synthesis blockade was expected in cells with DNA damage. However, the fact that undamaged cells were also unable to synthesise DNA in a sustained manner was unexpected and suggests that high dose AA directly affects cell metabolism. AA has been previously reported to decrease ATP levels in neuroblastoma cells [2], prostate cancer [3], ovarian cancer [4] pancreatic cancer [5]. This decrease in ATP was shown to be a result of genotoxic stress in the form of 8OH-dG lesions, which were repaired by PARP-1, leading to consumption of cytoplasmic NAD^+ (a cofactor of PARP-1). Decreased levels of NAD^+ inhibited glycolysis and glycolytic ATP production [2–5]. However, the cells that remained undamaged also stopped synthesising DNA and yet they have no need to activate PARP-1 with subsequent depletion of NAD^+ levels. We hypothesise that undamaged cells do not synthesise DNA because of depleted cellular ATP, due to direct oxidative damage to components of the energy metabolism pathways. In support of this hypothesis, Yun and colleagues recently showed that intracellular reduction of DHA to AA killed glycolysis-driven KRAS and BRAF mutated colorectal cancer cells through inhibition of glycolysis resulting in ATP depletion [6,7].

Intracellular NADH flux is a good measure of overall cellular energy metabolism. The tetrazolium dye MTS is reduced intracellularly (predominantly by NADH generated during glycolysis) as well as extracellularly (predominantly by NADH originating from the mitochondria) in the presence of an intermediate electron acceptor [30]. The strong decrease in MTS reduction at 1 h closely mimicked the strong drop in cellular ATP levels as well as a decrease in mitochondrial membrane potential. This may reflect both a lack of NAD^+ (required to fix 8OH-dG lesions) and/or a direct inhibition of glycolysis, Krebs cycle and MET activity, possibly due to oxidative damage to their components. In this respect it is of interest to note that H_2O_2 was previously shown to specifically damage the adenine nucleotide translocase (ANT) [32]. ANT is an inner mitochondrial membrane translocase which delivers ATP from the mitochondrial matrix to hexokinase II to facilitate the first step in glycolysis [33]. Oxidative damage to ANT inhibits glycolysis and thus glycolytic ATP production [32]. Declining glycolytic rates limit the amount of pyruvate entering the mitochondria which limits Krebs cycle activity, decreasing NADH and FADH2 levels that fuel MET, oxidative phosphorylation (OXPHOS), mitochondrial membrane potential and generate mitochondrial ATP. Superoxide is produced in the mitochondria during MET as a result of premature leakage of electrons at respiratory complexes I and III [34]. The small increase in superoxide levels we observed, combined with a decrease in membrane potential after AA and H_2O_2 treatments, suggests increased leakage due to oxidative damage to respiratory complexes. Both ATP level and MTS reduction 1 h after AA exposure were excellent predictors for cell survival 48 h later. Results presented in this paper show that both damaged and undamaged cells halt DNA synthesis within 2 h of exposure to AA which is not resolved 4 days later. This suggests that high dose AA and H_2O_2 generate both genotoxic and metabolic stress which contribute to blocking DNA synthesis in AA sensitive GBM cell lines. The specific contribution of each type of stress is likely to differ between cell lines and between cells of the same cell line.

The effect of H_2O_2 and high dose AA as mediators of genotoxic and metabolic stress were very similar in three of the four cell lines—T98G cells were less affected by AA than by H_2O_2 in all respects. This was expected, as this cell line is less sensitive to AA due to its high antioxidant capacity with an IC_{50} that is at least two times higher than that of the other cell lines [24]. It is possible that the antioxidant capacity of T98G was overwhelmed by a H_2O_2 bolus but able to cope with the H_2O_2 generated over a period of time from external AA.

5. Conclusions

This paper confirms that the mechanism of action of high dose AA is likely mediated by H_2O_2 generation as exposure to high dose AA and H_2O_2 abrogated DNA synthesis in cells with damaged and undamaged DNA. Both genotoxic stress and metabolic stress contributed to DNA synthesis arrest in DNA damaged cells. However, DNA synthesis arrest in undamaged cells can only be explained by direct oxidative damage to components of mitochondrial energy production.

Acknowledgments: This research was supported by Genesis Oncology Trust and the Neurological Foundation of New Zealand.

Author Contributions: Patries M. Herst, Melanie J. McConnell and Maria Leticia Castro conceived and designed the experiments; Maria Leticia Castro and Georgia M. Carson performed the experiments; Maria Leticia Castro and Patries M. Herst analyzed the data; Patries M. Herst wrote the manuscript with constructive feedback from Melanie J. McConnell, Maria Leticia Castro and Georgia M. Carson and all authors approved the final manuscript.

Conflicts of Interest: The authors declare no conflict of interest.

References

1. Parrow, N.L.; Leshin, J.A.; Levine, M. Parenteral ascorbate as a cancer therapeutic: A reassessment based on pharmacokinetics. *Antioxid. Redox Signal.* **2013**, *19*, 2141–2156. [CrossRef] [PubMed]
2. Bruchelt, G.; Schraufstätter, I.U.; Niethammer, D.; Cochrane, C. Ascorbic Acid Enhances the Effects of 6-Hydroxydopamine and H_2O_2 on Iron-dependent DNA Strand Breaks and Related Processes in the Neuroblastoma Cell Line SK-N-SH. *Cancer Res.* **1991**, *51*, 6066–6072.
3. Chen, P.; Yu, J.; Chalmers, B.; Drisko, J.; Yang, J.; Li, B.; Chen, Q. Pharmacological ascorbate induces cytotoxicity in prostate cancer cells through ATP depletion and induction of autophagy. *Anticancer Drugs* **2012**, *23*, 437–444. [CrossRef] [PubMed]
4. Ma, Y.; Chapman, J.; Levine, M.; Polireddy, K.; Drisko, J. High dose parenteral ascorbate enhanced chemosensitivity of ovarian cancer and reduced toxicity of chemotherapy. *Sci. Transl. Med.* **2014**, *6*, 222ra18. [CrossRef] [PubMed]
5. Du, J.; Martin, S.M.; Levine, M.; Wagner, B.A.; Buettner, G.R.; Wang, S.H.; Taghiyev, A.F.; Du, C.; Knudson, C.M.; Cullen, J.J. Mechanisms of Ascorbate-Induced Cytotoxicity in Pancreatic Cancer. *Clin. Cancer Res.* **2010**, *16*, 509–520. [CrossRef] [PubMed]
6. Yun, J.; Mullarky, E.; Lu, C.; Bosch, K.; Kavalier, A.; Rivera, K.; Roper, J.; Chio, I.I.C.; Giannopoulou, E.G.; Rago, C.; et al. Vitamin C selectively kills *KRAS* and *BRAF* mutant colorectal cancer cells by targeting GAPDH. *Science* **2015**, *350*, 1391–1396. [CrossRef] [PubMed]
7. Van der Reest, J.; Gottlieb, E. Anti-cancer effects of vitamin C revisited. *Cell Res.* **2016**, *26*, 269–270. [CrossRef] [PubMed]
8. Kuiper, C.; Dachs, G.U.; Munn, D.; Currie, M.J.; Robinson, B.A.; Pearson, J.F.; Vissers, M.C.M. Increased Tumor Ascorbate is Associated with Extended Disease-Free Survival and Decreased Hypoxia-Inducible Factor-1 Activation in Human Colorectal Cancer. *Front Oncol.* **2014**, *4*, 1–10. [CrossRef] [PubMed]
9. Kuiper, C.; Molenaar, I.G.M.; Dachs, G.U.; Currie, M.J.; Sykes, P.H.; Vissers, M.C.M. Low ascorbate levels are associated with increased hypoxia-inducible factor-1 activity and an aggressive tumor phenotype in endometrial cancer. *Cancer Res.* **2010**, *70*, 5749–5758. [CrossRef] [PubMed]
10. Bonuccelli, G.; De Francesco, E.M.; DeBoer, R.; Tanowitz, H.B.; Lisanti, M.P. NADH autofluorescence, a new metabolic biomarker for cancer stem cells: Identification of Vitamin C and CAPE as natural products targeting "stemness". *Oncotarget* **2017**, *8*, 20667–20678. [CrossRef] [PubMed]
11. Du, J.; Cullen, J.J.; Buettner, G.R. Ascorbic acid: Chemistry, biology and the treatment of cancer. *Buochim. Biophys. Acta* **2012**, *1826*, 443–457. [CrossRef] [PubMed]
12. Du, J.; Cieslak, J.A.; Welsh, J.L.; Sibenaller, Z.A.; Allen, B.G.; Wagner, B.A.; Kalen, A.L.; Doskey, C.M.; Strother, R.K.; Button, A.M.; et al. Pharmacological Ascorbate Radiosensitizes Pancreatic Cancer. *Cancer Res.* **2015**, *75*, 3314–3326. [CrossRef] [PubMed]

13. Chen, Q.; Espey, M.G.; Sun, A.Y.; Pooput, C.; Kirk, K.L.; Krishna, M.C.; Khosh, D.B.; Drisko, J.; Levine, M. Pharmacologic doses of ascorbate act as a prooxidant and decrease growth of aggressive tumor xenografts in mice. *Proc. Natl. Acad. Sci. USA* **2008**, *105*, 11105–11109. [CrossRef] [PubMed]

14. Monti, D.A.; Mitchell, E.; Bazzan, A.J.; Littman, S.; Zabrecky, G.; Yeo, C.J.; Pillai, M.V.; Newberg, A.B.; Deshmukh, S.; Levine, M. Phase I evaluation of intravenous ascorbic acid in combination with gemcitabine and erlotinib in patients with metastatic pancreatic cancer. *PLoS ONE* **2012**, *7*, e29794. [CrossRef] [PubMed]

15. Welsh, J.L.; Wagner, B.A.; van't Erve, T.J.; Zehr, P.S.; Berg, D.J.; Halfdanarson, T.R.; Yee, N.S.; Bodeker, K.L.; Du, J.; Roberts, L.J., II; et al. Pharmacological ascorbate with gemcitabine for the control of metastatic and node-positive pancreatic cancer (PACMAN): results from a phase I clinical trial. *Cancer Chemother. Pharmacol.* **2013**, *71*, 765–775. [CrossRef] [PubMed]

16. Deubzer, B.; Mayet, F.; Kuçi, Z.; Niewisch, M.; Merkel, G.; Handgretinger, R.; et al. H_2O_2-mediated cytotoxicity of pharmacologic ascorbate concentrations to neuroblastoma cells: potential role of lactate and ferritin. *Cell. Physiol. Biochem.* **2010**, *25*, 767–774. [CrossRef] [PubMed]

17. Duarte, T.L.; Almeida, G.M.; Jones, G.D.D. Investigation of the role of extracellular H_2O_2 and transition metal ions in the genotoxic action of ascorbic acid in cell culture models. *Toxicol. Lett.* **2007**, *170*, 57–65. [CrossRef] [PubMed]

18. Valko, M.; Rhodes, C.J.; Moncol, J.; Izakovic, M.; Mazur, M. Free radicals, metals and antioxidants in oxidative stress-induced cancer. *Chem. Biol. Interact.* **2006**, *160*, 1–40. [CrossRef] [PubMed]

19. Baader, S.L.; Bruchelt, G.; Carmine, T.C.; Lode, H.N.; Rieth, A.G.; Niethammer, D. Ascorbic acid mediated iron release from cellular ferritin and its relation to DNA strand break formation in neuroblastoma cells. *J. Cancer Res. Clin. Oncol.* **1994**, *120*, 415–421. [CrossRef] [PubMed]

20. Wouters, B.; Begg, A. Irradiation-induced damage and the DNA damage response. In *Basic Clinical Radiobiology*, 4th, ed.; Joiner, M.C., van der Kogel, A.J., Eds.; Hodder & Arnold: London, UK, 2009; pp. 11–41.

21. Bartkova, J.; Hamerlik, P.; Stockhausen, M.-T.; Ehrmann, J.; Hlobilkova, A.; Laursen, H.; Kalita, O.; Kolar, Z.; Poulsen, H.S.; Broholm, H.; et al. Replication stress and oxidative damage contribute to aberrant constitutive activation of DNA damage signalling in human gliomas. *Oncogene* **2010**, *29*, 5095–5102. [CrossRef] [PubMed]

22. Stupp, R.; Hegi, M.; Mason, W.; van den Bent, M.; Taphoorn, M.; Janzer, R.; Ludwin, S.K.; Allgeier, A.; Fisher, B.; Belanger, K.; et al. Effects of radiotherapy with concomitant and adjuvant temozolomide versus radiotherapy alone on survival in glioblastoma in a randomised phase III study: 5-year analysis of the EORTC-NCIC trial. *Lancet Oncol.* **2009**, *10*, 459–466. [CrossRef]

23. Herst, P.M.; Broadley, K.W.R.; Harper, J.L.; McConnell, M.J. Pharmacological concentrations of ascorbate radiosensitize glioblastoma multiforme primary cells by increasing oxidative DNA damage and inhibiting G2/M arrest. *Free Radic. Biol. Med.* **2012**, *52*, 1486–1493. [CrossRef] [PubMed]

24. Castro, M.; McConnell, M.; Herst, P. Radio-sensitisation by pharmacological ascorbate in glioblastoma multiforme cells, human glial cells and HUVECs depends on their antioxidant and DNA repair capabilities and is not cancer specific. *Free Radic. Biol. Med.* **2014**, *75*, 200–209. [CrossRef] [PubMed]

25. Katsube, T.; Mori, M.; Tsuji, H.; Shiomi, T.; Wang, B.; Liu, Q.; Nenoi, M.; Onoda, M. Most hydrogen peroxide-induced histone H2AX phosphorylation is mediated by ATR and is not dependent on DNA double-strand breaks. *J. Biochem.* **2014**, *156*, 85–95. [CrossRef] [PubMed]

26. Berniak, K.; Rybak, P.; Bernas, T.; Zarębski, M.; Biela, E.; Zhao, H.; Darzynkiewicz, Z.; Dobrucki, J.W. Relationship between DNA damage response, initiated by camptothecin or oxidative stress, and DNA replication, analyzed by quantitative 3D image analysis. *Cytom. A* **2013**, *83*, 913–924. [CrossRef] [PubMed]

27. Zhao, H.; Dobrucki, J.; Rybak, P.; Traganos, F.; Halicka, D.H.; Darzynkiewicz, Z. Induction of DNA damage signaling by oxidative stress in relation to DNA replication as detected using "click chemistry". *Cytom. A* **2011**, *79*, 897–902. [CrossRef] [PubMed]

28. De Feraudy, S.; Revet, I.; Bezrookove, V.; Feeney, L.; Cleaver, J.E. A minority of foci or pan-nuclear apoptotic staining of gammaH2AX in the S phase after UV damage contain DNA double-strand breaks. *Proc. Natl. Acad. Sci. USA* **2010**, *107*, 6870–6875. [CrossRef] [PubMed]

29. Broadley, K.W.R.; Hunn, M.K.; Farrand, K.J.; Price, K.M.; Grasso, C.; Miller, R.J.; Hermans, I.F.; McConnell, M.J. Side population is not necessary or sufficient for a cancer stem cell phenotype in gliobastoma multiforme. *Stem Cells* **2011**, *29*, 452–461. [CrossRef] [PubMed]

30. Berrdige, M.; Herst, P.; Tan, A. Tetrazolium dyes as tools in cell biology: New insights into their cellular reduction. *Biotechnol. Annu. Rev.* **2005**, *11*, 127–152.

31. Limsirichaikul, S.; Niimi, A.; Fawcett, H.; Lehmann, A.; Yamashita, S.; Ogi, T. A rapid non-radioactive technique for measurement of repair synthesis in primary human fibroblasts by incorporation of ethynyl deoxyuridine (EdU). *Nucleic Acids Res.* **2009**, *37*, 1–10. [CrossRef] [PubMed]
32. Yan, L.J.; Sohal, R.S. Mitochondrial adenine nucleotide translocase is modified oxidatively during aging. *Proc. Natl. Acad. Sci. USA* **1998**, *95*, 12896–12901. [CrossRef]
33. Cerella, C.; Dicato, M.; Diederich, M. Modulatory roles of glycolytic enzymes in cell death. *Biochem. Pharmacol.* **2014**, *92*, 22–30. [CrossRef] [PubMed]
34. Quinlan, C.L.; Perevoshchikova, I.V.; Hey-Mogensen, M.; Orr, A.L.; Brand, M.D. Sites of reactive oxygen species generation by mitochondria oxidizing different substrates. *Redox Biol.* **2013**, *1*, 304–312. [CrossRef] [PubMed]

antioxidants

MDPI

Article

No Reported Renal Stones with Intravenous Vitamin C Administration: A Prospective Case Series Study

Melissa Prier [1], Anitra C. Carr [2] and Nicola Baillie [3,*]

[1] Feedback Research Ltd., Auckland 1050, New Zealand; melissa@feedbackresearch.co.nz
[2] Department of Pathology and Biomedical Science, University of Otago, Christchurch 8011, New Zealand; anitra.carr@otago.ac.nz
[3] Integrated Health Options Ltd., Auckland 1050, New Zealand
* Correspondence: nicky@integratedhealthoptions.co.nz; Tel.: +64-9524-7745

Received: 3 April 2018; Accepted: 19 May 2018; Published: 21 May 2018

Abstract: A few cases associating high dose intravenous vitamin C (IVC) administration with renal stone formation have been reported in the literature, however, no long-term studies investigating IVC administration and reported renal stones have been carried out. Our aim was to measure the frequency of reported renal stones in patients receiving IVC therapy. We carried out a prospective case series study of 157 adult patients who commenced IVC therapy at Integrated Health Options clinic between 1 September 2011 and 31 August 2012, with follow-up for 12 months. Inquiries into the occurrence of renal stones were conducted at enrolment, 6 and 12 months, and renal function blood tests were conducted at enrolment, 4 weeks and every 12 weeks thereafter in a subgroup of patients. No renal stones were reported by any patients in the study, despite 8% of the patients having a history of renal stones. In addition, the majority of patients investigated had stable renal function during the study period as evidenced by little change in serum creatinine levels and estimated glomerular filtration rate (eGFR) following IVC. In conclusion, IVC therapy was not associated with patient-reported renal stones. Although not the primary focus of this study, it was also observed that there was no significant change in mean serum creatinine or eGFR for those who had follow-up renal function blood tests.

Keywords: vitamin C; intravenous vitamin C; renal stones; kidney stones; oxalate; renal function; creatinine; glomerular filtration rate

1. Introduction

Intravenous vitamin C (IVC) has been used for many decades by health care practitioners for a number of indications, particularly cancer and infection [1–3]. IVC has been shown to be effective at decreasing common cancer-related symptoms and chemotherapy side-effects, thus improving overall patient quality of life [4]. Furthermore, recent studies indicate that IVC can improve the outcomes of patients with severe infections and burns, and improve post-surgical recovery [5,6]. Pharmacokinetic studies have found that the peak concentration of vitamin C achievable in blood plasma following oral ingestion is about 220 μmol/L, whereas at least 15 mmol/L is achievable following IVC administration [7]. One of the proposed mechanisms of IVC is to target cancer cells via an indirect pro-oxidant mechanism relying on transition metal ion-dependent generation of hydrogen peroxide; vitamin C must be injected intravenously in order to achieve sufficiently high plasma vitamin C concentrations to facilitate this mechanism [2,8]. Intravenous administration may also be required to enhance diffusion of vitamin C into solid tumours with subsequent modulation of important cell signaling pathways [9,10]. Lower IVC doses tend to be used for patients with severe infection with the proposed mechanism of IVC being primarily support of organ and immune function [5,11].

IVC is generally considered to be safe with few adverse effects, however, it has been recommended that IVC be used with caution in patients with renal impairment or failure, or a history of renal oxalate stones [1,3]. Renal dysfunction impairs the kidney's ability to clear high doses of vitamin C from the circulation and several cases of IVC toxicity have been reported in patients with underlying renal dysfunction [12–15]. However, because haemodialysis and haemofiltration are able to remove vitamin C from the circulation [16], renal dysfunction is not necessarily contraindicated for IVC therapy in critically ill patients. In fact, there is some evidence that IVC may improve acute kidney injury in critically ill patients and also decrease renal toxicity in oncology patients receiving chemotherapy [17–19].

A minor product of vitamin C metabolism is oxalate and elevated urinary oxalate excretion has been reported after oral vitamin C intakes of 2 g/d, raising concerns associating vitamin C with increased risk of renal stones, which can potentially result in oxalate nephropathy [20,21]. Interestingly, neither of these studies reported a difference in vitamin C-associated oxalate generation between stone formers and non-stone formers. Robitaille et al. monitored oxalic acid secretion after IVC administration (doses of 0.2–1.5 g/kg body weight) and showed that less than 0.5% of very large IVC doses are excreted as oxalic acid in people with normal renal function [22]. Nevertheless, several cases of renal stone formation have been reported by practitioners administering IVC, although the type of stone was not always specified [1]. A clinical study in patients receiving continuous IVC for cancer also reported one case of a renal stone three weeks into the study, although renal function blood tests remained normal [23].

Published clinical studies on IVC administration typically only monitor outcomes for up to eight weeks, and the focus has not been on renal stone incidence or specifically renal function. There is a gap in the literature around assessment of the incidence of renal stones in patients receiving IVC over a longer time frame. There is minimal information on the time frame over which renal stones develop. Renal stones causing renal colic can develop in three months or less in high risk environments [24], although some may take longer than 12 months to develop [25,26]. Therefore, in the interests of patient safety, we sought to investigate whether IVC therapy was associated with renal stone presentation. The key parameters under investigation in this study were the patient-reported incidence of renal stones, considered reliable because of the inability to ignore renal colic which causes acute severe pain. Changes in serum creatinine and estimated glomerular filtration rate (eGFR) were also monitored in a subgroup of patients.

2. Materials and Methods

2.1. Study Design, Setting and Participants

This study was a prospective case series study which took place in a single private medical clinic, Integrated Health Options Ltd., in Auckland, New Zealand. The study received ethical approval from the Northern X Regional Ethics Committee (Reference NTX11/EXP/189).

Consecutive new adult patients seeking IVC therapy at the clinic were enrolled in the study between 1 September 2011 and 31 August 2012. To be included in the study of renal stone reporting, eligible participants were those: who had their first admission to the clinic during the study enrolment period, and received at least one IVC infusion, exclusively at the clinic, and had IVC as their main IV therapy. Exclusion criteria were patients receiving IV chelation therapy, as this may be a possible confounding factor. There were no other exclusion criteria as IVC is seen to be generally safe [1,27]. Patients were also closely monitored throughout the treatment to ensure safety (see follow-up section).

2.2. Study Size

The study size was determined by the number of patients who met eligibility criteria during the 12-month study period and who consented to be enrolled. Of the 168 patients who met the eligibility criteria for the renal stone reporting group, 157 were enrolled (93%). Reasons for non-enrolment could not be identified. Power calculations indicated that 125 patients would be required to achieve 80% power to detect significant changes in renal function.

2.3. Baseline Data

At consultation, prospective patients for whom IVC was indicated were asked if they had a history of renal stones. If the patients wished to proceed with IVC therapy, they were given a standard laboratory form and requested to visit their local phlebotomy lab for tests including glucose-6-phosphate dehydrogenase (G6PD) activity and renal function. Participant characteristics (including age, gender, history of renal stones, mean number and dose of IVC treatments, mean creatinine and eGFR), and their presenting conditions were recorded.

2.4. Treatment

IVC therapy involved diluting vitamin C (500 mg/mL Ascor L 500, McGuff Pharmaceuticals, Santa Ana, CA, USA) in 250–1000 mL sterile water depending on dose to maintain an appropriate osmolarity [3]. Doses started at 15–25 g per infusion and, depending on the medical condition, increased gradually over several infusions to typically between 50 and 100 g per infusion based on body-weight, plasma vitamin C concentration and patient tolerance. Two to three infusions per week plus daily oral vitamin C (1000–2000 mg/day) were recommended to maintain high plasma vitamin C levels.

2.5. Follow-Up

Renal function blood tests were routinely ordered for all patients prior to commencing IVC therapy and repeated after 4 weeks then every 12 weeks for as long as the patient continued to have IVC treatments at the clinic. Changes in renal function were managed in such a way that if the eGFR reduced 15–20%, renal function blood tests were repeated within 1–2 weeks, and if the eGFR reduced more than 20%, IVC was promptly ceased, the patient reviewed and any possible causes addressed [28,29]. Renal function tests were repeated 1–2 weeks later and if normalized or improving, IVC was restarted and monitored closely until stabilized. If levels continued to reduce, IVC was not restarted.

Enrolled patients were contacted by phone 6 months and 12 months after admission and asked a standardised question about experiencing, or seeking medical advice for, symptoms of renal stones or renal colic since starting IVC. If a patient died during the follow-up period, a coroner's report or death certificate was requested to check whether renal stones or renal impairment/failure was the cause of death or the antecedent cause of death. Due to a significant proportion of patients presenting with advanced cancer, a number of patients died during the study period (37/157, or 24%). Some patients were lost to follow-up due to moving overseas or lost to contact (three at six months, and six at 12 months).

2.6. Data Sources and Measurement

The patient-reported incidence of renal stones at the initial consult and follow-up phone interviews were recorded in a spreadsheet by a doctor. For those patients who died during the study period, information recorded on their death certificates was obtained from the Deputy Registrar General of Births, Deaths and Marriages. Other data collected by the treating physician included patient demographics, number of IVC infusions, renal function laboratory results and comments on reason for treatment cessation. Telephone interviews were conducted by the clinic nurses to investigate reported renal stones even after the patient stopped treatment. The last recorded renal function blood tests were used for comparison with their pre-IVC tests to measure changes in renal function.

2.7. Statistical Analysis

Statistical analyses were performed in Microsoft Excel 2007 (Microsoft Corporation, Redmond, WA, USA) with the Real Statistics 2.16.2 (Charles Zaiontz, http://www.real-statistics.com) data analysis tools add-on to conduct non-parametric tests. The Wilcoxon Signed-Rank test was used for paired samples (i.e., before and after), the Mann–Whitney test was used to compare two independent groups and the chi-squared test was used to compare groups with qualitative variables such as gender. A two-tailed *p*-value less than 0.05 was considered statistically significant.

3. Results

3.1. Participants

Of 168 new patients presenting to the clinic during the 12-month enrolment period, 157 (93%) eligible patients were enrolled in the study. At six months 154 (98%), and at 12 months 148 (94%) were able to be followed up by phone interviews, or death certificate information regarding reported renal stones. Approximately half (78/157) of the patients stopped IVC therapy before follow-up renal function blood tests were completed (i.e., within four weeks). Patient characteristics are presented in Table 1; it should be noted that 12 (8%) of the patients had a history of renal stones. The major presenting conditions were cancer and infectious diseases or requiring immune support (Table 2).

Table 1. Characteristics of participants.

Characteristic	Total Cohort [1]	No History of Renal Stones	History of Renal Stones	*p* Value [2]
N (%)	157 (100) [3]	141 (90)	12 (8)	
Mean age, years (range)	54 (17–86)	54 (17)	58 (12)	0.332
Male/female, *n*	62/95	53/88	9/3	0.038
Mean number of IVC treatments, n (range)	15 (1–119)	14 (19)	16 (13)	0.799
Mean IVC dose per session, g (range)	50 (15–125)	49 (23)	59 (24)	0.159
Mean creatinine pre-IVC, μmol/L (SD)	70 (20)	70 (21)	76 (18)	0.323
Mean eGFR pre-IVC, mL/min/1.73 m^2 (SD)	82 (12)	82 (14)	81 (14)	0.932

[1] Ethnicity: European 77%, Asian 8%, Maori/Pacific people 6%, Middle Eastern 1%, not stated 8%. [2] Comparisons between the two subgroups, unpaired *t* test used for continuous data, and chi-square test used for categorical data. [3] Separate data for patients with unknown history of renal stones are not shown (*n* = 4).

Table 2. Presenting conditions of participants.

Presenting Condition	Total Cohort *N* (%)	No History of Renal Stones	History of Renal Stones
Cancer	76 (48)	66 (47)	6 (50)
Infectious diseases and/or immune support	54 (34)	50 (35)	4 (33)
Neurological, musculoskeletal, and skin disorders	13 (8)	11 (8)	2 (17)
Fatigue	6 (4)	6 (4)	0
Other (e.g., pre-surgery, Crohn's disease, dental, fracture, haemorrhoids)	8 (5)	8 (6)	0

3.2. Outcome Data

3.2.1. Reported Renal Stones

Six months after commencing IVC therapy, 153 out of 157 patients (124 alive patients and 29 deceased patients) had no reported renal stones or no renal stones causing renal impairment/failure recorded on death certificates (Figure 1). One patient with pre-existing renal problems had died of renal failure (refer to Patient A case report below) and three patients had moved overseas and lost contact, so their renal stone status could not be determined. At 12 months, 118 (112 alive patients and six deceased patients) had no reported renal stones or no renal stones causing renal impairment/failure recorded on death certificates (Figure 1). Contact was lost with six patients (four patients had moved overseas, one was deceased, and failed to make contact with one), so their renal stone status could not be determined.

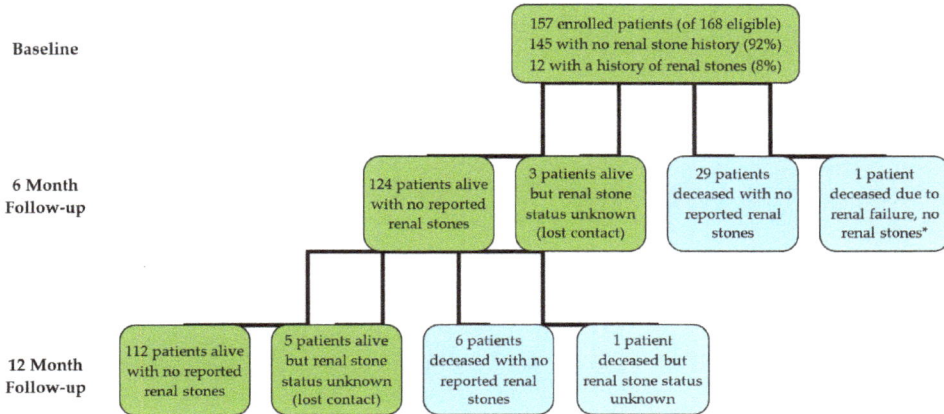

Figure 1. Flow diagram of participants involved in analysis of reported renal stones. * See Patient A case report.

3.2.2. Renal function tests

Paired data for renal function blood tests included 79 and 71 patient results for creatinine and eGFR, respectively. Forty-five (57%) of the patients showed an increase in serum creatinine concentrations, although only 3 of these went above normal (Table 3). The mean change in serum creatinine concentration was +1.8 ± 11.0 µmol/L (4.1 ± 15.7%). This was not statistically significant and was within reported biological variation expectations (4.7% in healthy individuals and 8.9% in people with renal impairment) [30]. Only 6 (8%) of the patients showed a decrease of ≥15% in eGFR (Table 3). The mean change in eGFR was −0.7 ± 9.9 mL/min/1.73 m^2 (0.1 ± 13.5%) and was not statistically significant. There was no significant relationship between gender (chi-squared p-value > 0.05).

Table 3. Changes in participant renal function.

Observation	Patient N (%)
Creatinine Increased [1]	45/79 (57)
Low to Normal	2/45 (4)
Normal to Normal	40/45 (89)
Normal to High	3/45 (7)
eGFR Decreased ≥ 15% [2]	6/71 (8)
Normal to Low	2/6 (33)
Low to Lower	4/6 (66)
Creatinine Increased + eGFR Decreased ≥ 15%	6/79 (9)

[1] Normal Creatinine: 60–105 µmol/L males, 45–90 µmol/L females. [2] Normal eGFR: ≥ 90 mL/min/1.73 m^2.

3.3. Patient A Case Report

Patient A died due to renal failure within six months of starting the study, but there were no reported renal stones, nor did she have a history of renal stones. The patient was a 60-year-old woman with metastatic breast cancer diagnosed four and a half years earlier who had undergone a mastectomy, axillary dissection, chemotherapy and hormone treatments. She had been admitted to hospital for pulmonary emboli, post-renal obstruction from ascites, peritoneal spread and gastric invasion five weeks prior to presenting at the clinic. Her renal function had improved following drainage of ascites. Four weeks prior to attending the clinic, she had started a course of chemotherapy with vinorelbine, had two treatments then two weeks off and further treatment was discontinued due to increased liver function test results.

After her consultation at the clinic she had three IVC treatments in the week following. She was readmitted to hospital 11 days after starting IVC with recurrence of post-renal obstruction and ascites. Renal function did not improve despite further drainage of ascites. She died four days later.

The possibility that IVC contributed to Patient A's renal failure cannot be excluded, but it is considered unlikely as she had proven recurrence of a pre-existing condition that had recently caused acute renal injury, namely post-renal obstruction secondary to cancer-related ascites. There were no renal stones reported from the investigations performed in hospital, or on her death certificate.

4. Discussion

The primary outcome of this prospective study was that no renal stones were reported by any patient who received IVC during the study period, despite 8% of participants reporting previous episodes of renal stones. This study showed that IVC was not associated with any cases of symptomatic renal stones or renal stones causing renal failure/death in 100% of patients able to be followed up over a 12-month period (94% follow-up). Although not the focus of this study, half of the eligible patients had follow-up creatinine and eGFR results available. Of the patients that provided both before and after renal function blood tests (50% of the patients), IVC therapy was not associated with a significant detrimental effect on renal function.

The purported risk of renal stones related to vitamin C has been based mainly on increased oxalate excretion rather than actual occurrence of renal stones [20,21,31], and issues around possible ex vivo formation of oxalate during sample storage and processing have been highlighted [22,32]. Large epidemiological studies have shown no association between vitamin C intake or status and increased prevalence of renal stones, particularly in women [33–36], although several studies showed increased risk among men [36–38]. Epidemiological studies can, however, only indicate associations, not causal relationships and are hampered by many confounders such as fluid intake and a lack of information on renal stone composition [36,38]. Furthermore, many rely upon self-reported food frequency questionnaires which are not necessarily a good indicator of plasma vitamin C status [39]. Simon et al. measured circulating vitamin C levels and found no association with renal stones in a large cohort of over ten thousand US adults [33].

There have been numerous case reports in the literature of oxalate nephropathy secondary to high dose oral vitamin C intake [40–49]. However, a number of these cases are complicated by dehydration due to diarrhoea and vomiting [42,43], or acute kidney injury and transplantation requiring dialysis [46–48]. Although no significant changes in oxalate metabolism were observed in a study involving ten end-stage renal patients who took 100 mg/day of vitamin C [50], higher doses have been associated with increased plasma oxalate levels in haemodialysis patients [51]. However, it should be noted that oxalate excretion may be increased in haemodialysis patients even without vitamin C administration [52].

Nevertheless, for those with severe renal impairment or at risk of severe renal impairment, such as in the case of Patient A, the warning to be cautious with regard to administering IVC therapy should be heeded [3]. In support of this, a previous case report that linked IVC with oxalate nephropathy involved a single infusion of 45 g IVC for a patient already diagnosed with a nephrotic syndrome [13]. Another case study involved a single 60 g IVC infusion of a patient with a metastatic carcinoma of the prostate and renal insufficiency [14]. Neither of the reports showed full renal function blood tests and both patients were renally impaired prior to IVC treatment. In the clinic where the current study was conducted, renal function blood tests are frequently monitored, especially in those patients who have renal impairment, and IVC would not commence in cases of severe renal impairment or be ceased if significant progressive deterioration in renal function occurred.

Ours is the first long-term study investigating IVC and reported renal stones. A strength of this study is that adult patients of various ages presented with a wide variety of conditions and were not excluded from the study based on their renal function prior to attending the clinic. They also received a wide range of IVC doses (from 1 g to 119 g). Therefore, the study population may be considered representative of adults receiving IVC therapy in a clinical setting. Although the incidence of renal stones in the general population is only 1–3/1000/year [53], our cohort consisted of 12 participants with a history of renal stones, which can increase the risk of developing recurrent stones by up to 50% [54,55]. Despite this increased risk and the high doses of IVC being administered, which require clearance by the kidneys, we did not observe any recurrent renal stones in these participants over the duration of the study.

There were a number of limitations to this study. There was a potential for self-selection bias as patients could choose to stop IVC therapy or not undergo scheduled renal function blood tests. Although only 6% were lost to follow-up, 11 eligible patients were not enrolled for reasons that could not be identified. There were also confounding factors such as pre-existing disease and complications or co-morbidities, other concurrent or recent therapies and lifestyle factors such as diet. This study relied on patient-reported renal stones and for those patients who died, information was taken from their death certificates. Although self-report of renal stones has been validated [56,57], the study did not include radiographic investigations of the participants so would not have detected asymptomatic renal stones [58]. The study was also not sufficiently powered to make statistically significant conclusions regarding renal function. Furthermore, in relation to the timeframe of the study relative to renal stone development, some renal stones may take longer than 12 months to develop/cause symptoms [25,26], in which case they would not have been included in this study.

5. Conclusions

No renal stones were reported by patients or recorded on death certificates in 157 patients receiving IVC therapy during a 12-month period, despite 8% of the patients having a history of renal stones. The majority of patients had stable renal function and our data suggests that changes in eGFR were negligible on average, but due to statistical weaknesses (numbers lost to follow-up in this group) these findings cannot rule out the possibility that IVC therapy may contribute to renal dysfunction. Therefore, further research is required with larger sample sizes, longer term follow-up (>12 months) and radiographic investigations.

Author Contributions: N.B. recruited patients and collected data, M.P. analysed data, A.C., M.P. and N.B. wrote the paper.

Acknowledgments: We thank all of the patients who participated in this study. We also acknowledge Selene Peng for assisting with data analysis. A.C. is supported by a Health Research Council of New Zealand Sir Charles Hercus Health Research Fellowship (#16/037).

Conflicts of Interest: The authors declare no conflict of interest.

References

1. Padayatty, S.J.; Sun, A.Y.; Chen, Q.; Espey, M.G.; Drisko, J.; Levine, M. Vitamin C: Intravenous use by complementary and alternative medicine practitioners and adverse effects. *PLoS ONE* **2010**, *5*, e11414. [CrossRef] [PubMed]
2. Parrow, N.L.; Leshin, J.A.; Levine, M. Parenteral ascorbate as a cancer therapeutic: A reassessment based on pharmacokinetics. *Antioxid. Redox Signal* **2013**, *19*, 2141–2156. [CrossRef] [PubMed]
3. Mikirova, N.A.; Casciari, J.J.; Hunninghake, R.E.; Riordan, N.H. Intravenous ascorbic acid protocol for cancer patients: Scientific rationale, pharmacology, and clinical experience. *Funct. Foods Health Dis.* **2013**, *3*, 344–366.
4. Carr, A.C.; Vissers, M.C.M.; Cook, J.S. The effect of intravenous vitamin C on cancer- and chemotherapy-related fatigue and quality of life. *Front. Oncol.* **2014**, *4*, 1–7. [CrossRef] [PubMed]

5. Carr, A.C.; Shaw, G.M.; Fowler, A.A.; Natarajan, R. Ascorbate-dependent vasopressor synthesis: A rationale for vitamin C administration in severe sepsis and septic shock? *Crit. Care* **2015**, *19*, e418. [CrossRef] [PubMed]

6. Fukushima, R.; Yamazaki, E. Vitamin C requirement in surgical patients. *Curr. Opin. Clin. Nutr. Metab. Care* **2010**, *13*, 669–676. [CrossRef] [PubMed]

7. Padayatty, S.J.; Sun, H.; Wang, Y.; Riordan, H.D.; Hewitt, S.M.; Katz, A.; Wesley, R.A.; Levine, M. Vitamin C pharmacokinetics: Implications for oral and intravenous use. *Ann. Intern. Med.* **2004**, *140*, 533–537. [CrossRef] [PubMed]

8. Mastrangelo, D.; Pelosi, E.; Castelli, G.; Lo-Coco, F.; Testa, U. Mechanisms of anti-cancer effects of ascorbate: Cytotoxic activity and epigenetic modulation. *Blood Cells Mol. Dis.* **2017**, *69*, 57–64. [CrossRef] [PubMed]

9. Kuiper, C.; Vissers, M.C.; Hicks, K.O. Pharmacokinetic modeling of ascorbate diffusion through normal and tumor tissue. *Free Radic. Biol Med.* **2014**, *77*, 340–352. [CrossRef] [PubMed]

10. Kuiper, C.; Vissers, M.C. Ascorbate as a co-factor for Fe- and 2-oxoglutarate dependent dioxygenases: Physiological activity in tumor growth and progression. *Front. Oncol.* **2014**, *4*, 359. [CrossRef] [PubMed]

11. Carr, A.C.; Maggini, S. Vitamin C and immune function. *Nutrients* **2017**, *9*, 1211. [CrossRef] [PubMed]

12. McAllister, C.J.; Scowden, E.B.; Dewberry, F.L.; Richman, A. Renal failure secondary to massive infusion of vitamin C. *JAMA* **1984**, *252*, 1684. [CrossRef] [PubMed]

13. Lawton, J.M.; Conway, L.T.; Crosson, J.T.; Smith, C.L.; Abraham, P.A. Acute oxalate nephropathy after massive ascorbic acid administration. *Arch. Intern. Med.* **1985**, *145*, 950–951. [CrossRef] [PubMed]

14. Wong, K.; Thomson, C.; Bailey, R.R.; McDiarmid, S.; Gardner, J. Acute oxalate nephropathy after a massive intravenous dose of vitamin C. *Aust. N. Z. J. Med.* **1994**, *24*, 410–411. [CrossRef] [PubMed]

15. Cossey, L.N.; Rahim, F.; Larsen, C.P. Oxalate nephropathy and intravenous vitamin C. *Am. J. Kidney Dis.* **2013**, *61*, 1032–1035. [CrossRef] [PubMed]

16. Story, D.A.; Ronco, C.; Bellomo, R. Trace element and vitamin concentrations and losses in critically ill patients treated with continuous venovenous hemofiltration. *Crit. Care Med.* **1999**, *27*, 220–223. [CrossRef] [PubMed]

17. Marik, P.E.; Khangoora, V.; Rivera, R.; Hooper, M.H.; Catravas, J. Hydrocortisone, vitamin C, and thiamine for the treatment of severe sepsis and septic shock: A retrospective before-after study. *Chest* **2017**, *151*, 1229–1238. [CrossRef] [PubMed]

18. Dennis, J.M.; Witting, P.K. Protective role for antioxidants in acute kidney disease. *Nutrients* **2017**, *9*, 718. [CrossRef] [PubMed]

19. Ma, Y.; Chapman, J.; Levine, M.; Polireddy, K.; Drisko, J.; Chen, Q. High-dose parenteral ascorbate enhanced chemosensitivity of ovarian cancer and reduced toxicity of chemotherapy. *Sci. Transl. Med.* **2014**, *6*, 222ra18. [CrossRef] [PubMed]

20. Massey, L.K.; Liebman, M.; Kynast-Gales, S.A. Ascorbate increases human oxaluria and kidney stone risk. *J. Nutr.* **2005**, *135*, 1673–1677. [CrossRef] [PubMed]

21. Traxer, O.; Huet, B.; Poindexter, J.; Pak, C.Y.; Pearle, M.S. Effect of ascorbic acid consumption on urinary stone risk factors. *J. Urol.* **2003**, *170 Pt 1*, 397–401. [CrossRef] [PubMed]

22. Robitaille, L.; Mamer, O.A.; Miller, W.H., Jr.; Levine, M.; Assouline, S.; Melnychuk, D.; Rousseau, C.; Hoffer, L.J. Oxalic acid excretion after intravenous ascorbic acid administration. *Metabolism* **2009**, *58*, 263–269. [CrossRef] [PubMed]

23. Riordan, H.D.; Casciari, J.J.; Gonzalez, M.J.; Riordan, N.H.; Miranda-Massari, J.R.; Taylor, P.; Jackson, J.A. A pilot clinical study of continuous intravenous ascorbate in terminal cancer patients. *P. R. Health Sci. J.* **2005**, *24*, 269–276. [PubMed]

24. Evans, K.; Costabile, R.A. Time to development of symptomatic urinary calculi in a high risk environment. *J. Urol.* **2005**, *173*, 858–861. [CrossRef] [PubMed]

25. Shepard, C.L.; Wang, G.; Hopson, B.D.; Bunt, E.B.; Assimos, D.G. Urinary tract stone development in patients with myelodysplasia subjected to augmentation cystoplasty. *Rev. Urol.* **2017**, *19*, 11–15. [PubMed]

26. Kok, D.J.; Boellaard, W.; Ridwan, Y.; Levchenko, V.A. Timelines of the "free-particle" and "fixed-particle" models of stone-formation: Theoretical and experimental investigations. *Urolithiasis* **2017**, *45*, 33–41. [CrossRef] [PubMed]

27. Stephenson, C.M.; Levin, R.D.; Spector, T.; Lis, C.G. Phase I clinical trial to evaluate the safety, tolerability, and pharmacokinetics of high-dose intravenous ascorbic acid in patients with advanced cancer. *Cancer Chemother. Pharmacol.* **2013**, *72*, 139–146. [CrossRef] [PubMed]

28. Ministry of Health. *Managing Chronic Kidney Disease in Primary Care: A National Consensus Statement*; Ministry of Health: Wellington, New Zealand, 2015.
29. Chronic Kidney Disease (CKD): Management in General Practice. 2014. Available online: https://www.kidneys.co.nz/resources/file/ckd_management_in_general_practice._2014_version.pdf (accessed on 16 March 2018).
30. Reinhard, M.; Erlandsen, E.J.; Randers, E. Biological variation of cystatin C and creatinine. *Scand. J. Clin. Lab. Investig.* **2009**, *69*, 831–836. [CrossRef] [PubMed]
31. Auer, B.L.; Auer, D.; Rodgers, A.L. Relative hyperoxaluria, crystalluria and haematuria after megadose ingestion of vitamin C. *Eur. J. Clin. Investig.* **1998**, *28*, 695–700. [CrossRef]
32. Gerster, H. No contribution of ascorbic acid to renal calcium oxalate stones. *Ann. Nutr. Metab.* **1997**, *41*, 269–282. [CrossRef] [PubMed]
33. Simon, J.A.; Hudes, E.S. Relation of serum ascorbic acid to serum vitamin B12, serum ferritin, and kidney stones in US adults. *Arch. Intern. Med.* **1999**, *159*, 619–624. [CrossRef] [PubMed]
34. Curhan, G.C.; Willett, W.C.; Rimm, E.B.; Stampfer, M.J. A prospective study of the intake of vitamins C and B6, and the risk of kidney stones in men. *J. Urol.* **1996**, *155*, 1847–1851. [CrossRef]
35. Curhan, G.C.; Willett, W.C.; Speizer, F.E.; Stampfer, M.J. Intake of vitamins B6 and C and the risk of kidney stones in women. *J. Am. Soc. Nephrol.* **1999**, *10*, 840–845. [PubMed]
36. Ferraro, P.M.; Curhan, G.C.; Gambaro, G.; Taylor, E.N. Total, dietary, and supplemental vitamin C intake and risk of incident kidney stones. *Am. J. Kidney Dis.* **2016**, *67*, 400–407. [CrossRef] [PubMed]
37. Thomas, L.D.; Elinder, C.G.; Tiselius, H.G.; Wolk, A.; Akesson, A. Ascorbic acid supplements and kidney stone incidence among men: A prospective study. *JAMA Intern. Med.* **2013**, *173*, 386–388. [CrossRef] [PubMed]
38. Taylor, E.N.; Stampfer, M.J.; Curhan, G.C. Dietary factors and the risk of incident kidney stones in men: New insights after 14 years of follow-up. *J. Am. Soc. Nephrol.* **2004**, *15*, 3225–3232. [CrossRef] [PubMed]
39. Pullar, J.M.; Carr, A.C.; Vissers, M.C. Vitamin C supplementation and kidney stone risk. *N. Z. Med. J.* **2013**, *126*, 133–134. [PubMed]
40. Lamarche, J.; Nair, R.; Peguero, A.; Courville, C. Vitamin C-induced oxalate nephropathy. *Int. J. Nephrol.* **2011**, *2011*, 146927. [CrossRef] [PubMed]
41. McHugh, G.J.; Graber, M.L.; Freebairn, R.C. Fatal vitamin C-associated acute renal failure. *Anaesth Intensive Care* **2008**, *36*, 585–588. [PubMed]
42. Rathi, S.; Kern, W.; Lau, K. Vitamin C-induced hyperoxaluria causing reversible tubulointerstitial nephritis and chronic renal failure: A case report. *J. Med. Case Rep.* **2007**, *1*, 155. [CrossRef] [PubMed]
43. Nasr, S.H.; Kashtanova, Y.; Levchuk, V.; Markowitz, G.S. Secondary oxalosis due to excess vitamin C intake. *Kidney Int.* **2006**, *70*, 1672. [CrossRef] [PubMed]
44. Poulin, L.D.; Riopel, J.; Castonguay, V.; Mac-Way, F. Acute oxalate nephropathy induced by oral high-dose vitamin C alternative treatment. *Clin. Kidney J.* **2014**, *7*, 218. [CrossRef] [PubMed]
45. Gurm, H.; Sheta, M.A.; Nivera, N.; Tunkel, A. Vitamin C-induced oxalate nephropathy: A case report. *J. Community Hosp. Intern. Med. Perspect.* **2012**, *2*. [CrossRef] [PubMed]
46. Nankivell, B.J.; Murali, K.M. Images in clinical medicine. Renal failure from vitamin C after transplantation. *N. Engl. J. Med.* **2008**, *358*, e4. [CrossRef] [PubMed]
47. Yaich, S.; Chaabouni, Y.; Charfeddine, K.; Zaghdane, S.; Kharrat, M.; Kammoun, K.; Makni, S.; Boudawara, T.; Hachicha, J. Secondary oxalosis due to excess vitamin C intake: A cause of graft loss in a renal transplant recipient. *Saudi J. Kidney Dis. Transpl.* **2014**, *25*, 113–116. [CrossRef] [PubMed]
48. Colliou, E.; Mari, A.; Delas, A.; Delarche, A.; Faguer, S. Oxalate nephropathy following vitamin C intake within intensive care unit. *Clin. Nephrol.* **2017**, *88*, 354–358. [CrossRef] [PubMed]
49. Mashour, S.; Turner, J.F., Jr.; Merrell, R. Acute renal failure, oxalosis, and vitamin C supplementation: A case report and review of the literature. *Chest* **2000**, *118*, 561–563. [CrossRef] [PubMed]
50. Morgan, S.H.; Maher, E.R.; Purkiss, P.; Watts, R.W.; Curtis, J.R. Oxalate metabolism in end-stage renal disease: The effect of ascorbic acid and pyridoxine. *Nephrol. Dial Transplant.* **1988**, *3*, 28–32. [PubMed]
51. Canavese, C.; Petrarulo, M.; Massarenti, P.; Berutti, S.; Fenoglio, R.; Pauletto, D.; Lanfranco, G.; Bergamo, D.; Sandri, L.; Marangella, M. Long-term, low-dose, intravenous vitamin C leads to plasma calcium oxalate supersaturation in hemodialysis patients. *Am. J. Kidney Dis.* **2005**, *45*, 540–549. [CrossRef] [PubMed]

52. Costello, J.F.; Sadovnic, M.J.; Cottington, E.M. Plasma oxalate levels rise in hemodialysis patients despite increased oxalate removal. *J. Am. Soc. Nephrol.* **1991**, *1*, 1289–1298. [PubMed]

53. Curhan, G.C. Epidemiology of stone disease. *Urol. Clin. N. Am.* **2007**, *34*, 287–293. [CrossRef] [PubMed]

54. Lee, Y.H.; Huang, W.C.; Lu, C.M.; Tsai, J.Y.; Huang, J.K. Stone recurrence predictive score (SRPS) for patients with calcium oxalate stones. *J. Urol.* **2003**, *170 Pt 1*, 404–407. [CrossRef] [PubMed]

55. Ljunghall, S.; Danielson, B.G. A prospective study of renal stone recurrences. *Br. J. Urol.* **1984**, *56*, 122–124. [CrossRef] [PubMed]

56. Curhan, G.C.; Willett, W.C.; Rimm, E.B.; Stampfer, M.J. A prospective study of dietary calcium and other nutrients and the risk of symptomatic kidney stones. *N. Engl. J. Med.* **1993**, *328*, 833–838. [CrossRef] [PubMed]

57. Curhan, G.C.; Willett, W.C.; Speizer, F.E.; Spiegelman, D.; Stampfer, M.J. Comparison of dietary calcium with supplemental calcium and other nutrients as factors affecting the risk for kidney stones in women. *Ann. Intern. Med.* **1997**, *126*, 497–504. [CrossRef] [PubMed]

58. Bansal, A.D.; Hui, J.; Goldfarb, D.S. Asymptomatic nephrolithiasis detected by ultrasound. *Clin. J. Am. Soc. Nephrol.* **2009**, *4*, 680–684. [CrossRef] [PubMed]

![antioxidants logo] *antioxidants*

MDPI

Meeting Report

Symposium on Vitamin C, 15th September 2017; Part of the Linus Pauling Institute's 9th International Conference on Diet and Optimum Health

Anitra C. Carr

Department of Pathology, University of Otago, Christchurch, PO Box 4345, Christchurch 8140, New Zealand; anitra.carr@otago.ac.nz; Tel.: +64-3-364-0649

Received: 12 November 2017; Accepted: 18 November 2017; Published: 21 November 2017

1. Preface

The Linus Pauling Institute's 9th International Conference on Diet and Optimum Health took place on 13–15 September 2017 in Corvallis, OR, USA, on the beautiful Oregon State University campus. The theme was "Innovative approaches to improving health", and the event took place on the 100th anniversary of Linus Pauling joining Oregon State University as an undergraduate, and the 20th anniversary of the Linus Pauling Institute's move to Oregon State from Stanford University. The event also included a tribute to former LPI Director Balz Frei on his retirement (1997–2016). Dr. Frei received the Linus Pauling Institute Prize for Health Research, which is awarded in recognition of innovation and excellence in research relating to the roles of vitamins, essential minerals, and phytochemicals in promoting optimum health and preventing or treating disease, and the roles of oxidative/nitrative stress and antioxidants in human health and disease. The prize also recognizes successful efforts to disseminate and implement knowledge on diet, lifestyle, and health to enhance public health and reduce suffering from disease. I retain fond memories of being the first Postdoctoral Fellow at LPI under Balz Frei's mentorship (1998–2001). As such, it was my pleasure to convene a special one-day Symposium on Vitamin C (15th September 2017) as a tribute to Balz Frei and the legacy of Linus Pauling. Aside from being awarded two unshared Nobel Prizes, Linus Pauling is perhaps best known for his pioneering work in the 1970s on the role of vitamin C in cancer and infection, so it was fitting that the theme for the symposium was cancer and severe infection (sepsis). Huge breakthroughs have been made in these two fields since Linus Pauling first popularized them. The following abstracts are from the researchers and clinicians who were invited to present at this special symposium on vitamin C, with a focus on mechanisms and therapeutic use in cancer and sepsis.

2. Mechanisms of Vitamin C in Cancer

2.1. Ascorbic Acid Physiology and Pharmacology: Foundations for Cancer and Sepsis Therapies

Mark Levine, MD

National Institute of Diabetes and Digestive and Kidney Diseases, National Institutes of Health, Bethesda, MD 20892, USA; markl@bdg8.niddk.nih.gov

Recommended intakes for many vitamins are based on preventing deficiency, with a safety margin. Our hypotheses are that vitamin recommendations can and should be based on detailed, state-of-the-art physiological investigations in humans, using the clinical tools of concentration-function relationships and pharmacokinetics, coupled with cell biology and genomics. Clinical investigation tools were used to characterize vitamin C physiology in healthy men and women, ages 18–28 years. Utilizing a depletion-repletion design, vitamin C concentrations were found to be tightly controlled in plasma and cells, over a dose range of 30 to 2500 mg. Tight control had at least 4 components: bioavailability, or intestinal absorption; tissue transport; renal filtration, or renal reabsorption/excretion; and utilization. The first three components were regulated by two tissue transporters, mediated by identified transporters SLC23A1 (SVCT1) and SLC23A2 (SVCT2), as well as by as yet unidentified transporters. Recent studies indicate that red blood cells are a unique tissue compartment for ascorbate. In contrast to other tissues, red blood cells transport dehydroascorbic acid (oxidized ascorbate) utilizing the glucose transporter GLUT1. Bioavailability studies showed that percent intestinal absorption decreased as doses increased. Intravenous administration of doses above 100 mg produced ascorbate plasma concentrations that could not be achieved with oral dosing. These data indicated that, depending on dose and rate of administration, intravenous ascorbate produced pharmacologic plasma concentrations, with renal filtration restoring homeostasis.

Pharmacologic intravenously administered ascorbate, but not oral ascorbate, had the potential to decrease cancer growth in humans. Pharmacologic ascorbate was cytotoxic to cancer, but not to normal cells in vitro, both in animals and in small—but encouraging—studies in humans. Pharmacologic ascorbate mediated cancer cell death by generation of extracellular hydrogen peroxide (H_2O_2) in vivo. Pharmacologic ascorbate can be considered a pro-drug for delivery of pharmacologic H_2O_2 concentrations to the extracellular space. There is an ever-increasing multiplicity of downstream mechanisms of ascorbate-mediated cancer cell cytotoxicity that are H_2O_2-dependent. In specifically modified cell lines, mechanisms have been proposed that are H_2O_2-independent; for example, based on dehydroascorbic acid. However, this mechanism does not appear to have general applicability. Clinically, pharmacologic ascorbate has a surprisingly strong safety profile. Non-specificity, or promiscuity, of many oncology therapeutics is often harmful because of collateral damage to normal tissues in humans. In contrast, benefit is provided to patients by the promiscuity of pharmacologic ascorbate because of its safety and potential efficacy. Accelerated ascorbate utilization occurs in critically ill patients. Using similar pharmacokinetics principles as for cancer treatment, intravenous ascorbate has shown recent promise in treatment of sepsis. Based on transporter principles and red blood cell physiology, ascorbate either orally or intravenously has additional promise in delaying complications of diabetic microvascular disease. Considered together, the data indicate that exhaustive characterization of vitamin physiology in healthy people serves as a gateway to advances in disease treatment and prevention. Applying this approach to other vitamins will provide physiologic bases for vitamin recommendations, and may reveal unanticipated application to disease treatment.

2.2. Vitamin C-Dependent Regulation of the Hypoxic Response in Cancer

Margreet C.M. Vissers, PhD

Department of Pathology, University of Otago, Christchurch 8140, New Zealand; margreet.vissers@otago.ac.nz

Rapid tumor growth initiates hypoxic stress and activates the transcription factor hypoxia-inducible factor (HIF)-1, promoting angiogenesis, glycolysis and enhanced resistance to radio- and chemotherapy. HIF-1 is down-regulated by oxygen-sensing hydroxylases that require vitamin C (ascorbate) as cofactor and we have observed an inverse correlation between cell ascorbate levels and HIF-1 activity. Our hypothesis is that poor vascular ascorbate delivery in a growing tumor limits cell ascorbate content and augments the hypoxic response, thereby driving tumor growth. Raising tumor ascorbate could reverse this effect. Our in vitro modelling of ascorbate uptake into tissues is consistent with this hypothesis and suggests that supra-physiological concentrations are required to saturate tumor tissue.

We have shown that HIF-1 is moderated by intracellular ascorbate in tumors grown in the $Gulo^{-/-}$ mouse, a model of human vitamin C dependency. Pre-clinical studies with tumor tissue from cancer patients have also shown a correlation between cellular ascorbate levels, HIF-1 activity, tumor size and patient outcome in breast, colorectal, endometrial and renal cancers. These results suggest that raising tumor ascorbate could slow tumor growth by moderating HIF-1 activation. Our data from tumor-bearing Gulo$^{-/-}$ mice suggests that achieving mM plasma ascorbate levels by daily administration of supra-physiological doses exerts an anti-tumor effect, with decreased HIF-1 and vascular endothelial growth factor protein expression as well as reduced microvessel density, tumor hypoxia and tumor growth.

We have recently completed a pilot study with colorectal cancer patients who were given a high dose of vitamin C (1 g/kg) for four days prior to surgical removal of the tumor. Our analysis of the tumor tissue indicates that a significant increase in ascorbate levels is achieved through the high dose intervention that could impact on HIF-1 activation. Together, our results support the suppression of HIF-1 as an anti-tumor activity of ascorbate that may be useful in a clinical setting.

2.3. Vitamin C as a Multi-Targeting Anti-Cancer Agent

Qi Chen, PhD

Department of Pharmacology, Toxicology and Therapeutics, University of Kansas Medical Center, Kansas City, KS 66160, USA; qchen@kumc.edu

High-dose intravenous ascorbate (IVC) has attracted increasing interests as a low-toxic cancer therapy. IVC bypasses bioavailability barriers of oral ingestion, provides pharmacologic concentrations in tissues, and exhibits selective cytotoxic effects in cancer cells through peroxide formation. The selectivity is related to the mechanisms of action. We postulate that ascorbate-induced reactive oxygen species (ROS) has multiple mechanisms of action that preferably influence cancer cells. First, ascorbate-generated ROS induces DNA damage. Downstream to DNA damage, cellular NAD$^+$ decreases as an effect of poly ADP ribose polymerase (PARP) activation. Decrease of NAD$^+$ inhibited glyceraldehyde 3-phosphate dehydrogenase (GAPDH) activity and depletes ATP in cancer cells, while normal cells maintain their ATP levels. This phenomenon has its root in dysregulated glucose metabolism in cancer cells, known as the Warburg Effect, whereby cancer cells depend on a larger proportion on glycolysis for ATP, whereas normal cells depend more on oxidative phosphorylation. Second, lack of NAD$^+$ inhibits activity of Sirt-2, a tubulin deacetylase, and therefore increases tubulin acetylation, which in turn disrupts dynamics of microtubules. This influences cancer cells that are actively undergoing mitosis and migration. Third, when PARP is inhibited, excessive DNA damage results in cell death. Further, ascorbate inhibited epithelial-mesenchymal transition (EMT), an important process contributing to cancer metastasis. Finally, ascorbate enhanced collagen synthesis in tumor stroma. Despite the controversial reports on the effect of elevated collagen in tumor progression, the increased collagen by ascorbate treatment is associated with restriction of tumor invasion in our animal experiment and in patient.

Taken together, these data show multi-targeting effects of ascorbate that favor death/inhibition in cancer cells relative to normal cells. With minimal toxicity, the multi-targeting mechanism of ascorbate is advantageous because it could decrease the likelihood of resistance, and provides multiple opportunities for combination with standard chemo and radiation therapies.

2.4. Vitamin C and Bone Marrow Stem Cell Transplantation

Gerard M.J. Bos, MD, PhD

Maastricht University Medical Center, 6202 AZ Maastricht, The Netherlands; gerard.bos@mumc.nl

Bone marrow stem cell transplantation is used in the treatment of patients with cancer, mostly hematological types of cancers. Both autologous transplants and allogeneic transplants can be used. Autologous transplantation is used to recover bone marrow after high-dose chemotherapy, with or without radiotherapy. Allogeneic transplantation is used to induce a graft-versus-tumor response against a patient's malignant cells, based on an immunological mismatch between donor and recipient. Both treatments do induce a time period of immunosuppression in which the patient is at risk of infections. In autologous transplants, this risk is relatively low (mortality 2–5%). In allogeneic transplants, this risk is higher, and depends partly on the source of the transplant, potentially leading to a state of immune-deficiency of over 3 months.

This clinical question led us to study the possible option of adding back of T and/or Natural Killer (NK) cells cultured from stem cells in vitro. In vitro, it is possible to culture (pre-) T cells from stem cells in the presence of feeder cells transfected with Notch-Ligands, which are crucial for T cell development. In order to test a feeder-free system for GMP-accredited clinical use, we observed that different media gave different results; the difference could be explained by the presence or absence of vitamin C. In the presence of vitamin C, double-positive T cells from stem cells can be obtained in the absence of stromal cells [1]. However, for further development to single positive T cells, stromal cells are still needed. In addition, we also observed that NK cells derived and matured faster from stem cells, as well as from early pre T/NK cells, in the presence of vitamin C [2]. These cells are completely functional NK cells.

Based on these in vitro data, we went back to the clinic to see if patients that are treated for hematological diseases, with or without a stem cell transplantation, might have low vitamin C levels. Patients had lower vitamin C levels (20.4 microMol/L; N = 42) compared to 65 microMol/L in controls [3]. 20% of the patients had levels that were below 11 microMol/L or were undetectable, and were considered to be vitamin C deficient. In a more recent prospective study, we confirm this observation; patients have lower vitamin C levels while on treatment, compared to family members as controls.

These data are the basis for an intervention study using vitamin C to see if leukocyte recovery, and therefore infections, morbidity and mortality, can be improved after stem-cell transplantation for hematological malignancies.

3. Vitamin C Therapy in Cancer

3.1. The Chemical Biology of IV-C in Cancer Treatment: Basic Science to Clinical Trials

Garry R. Buettner, PhD

Department of Radiation Oncology, The University of Iowa, Iowa City, IA 52242-1089, USA; garry-buettner@uiowa.edu

Ascorbate functions as a versatile reducing agent. At pharmacological doses (P-AscH$^-$, [plasma] $\approx$ 20 mM), achievable through intravenous delivery, oxidation of AscH$^-$ can produce a high flux of H_2O_2 in tumors. Normal cells/tissues seem not to be affected by an increased flux of H_2O_2, while exposure to an increased flux of H_2O_2 is detrimental to many cancer cells. I will address three basic issues for the use of P-AscH$^-$ in the treatment of cancer: (1) the oxidation of ascorbate to produce a flux of H_2O_2; (2) catalase as the first-line defense of cells against an increased flux of H_2O_2; and (3) potential downstream modulators of the cellular response to this oxidative challenge. Our laboratory is quantitatively addressing these three aspects of the potential use of P-AscH$^-$ to treat cancer. I will show how data from experiments that address these issues are used to guide clinical trials.

The encouraging data from our basic science efforts have underpinned six clinical trials on the use of P-AscH$^-$ as an adjuvant to the standard of care at The University of Iowa; one completed (Phase 1, pancreatic cancer with gemcitabine); one terminated (Phase 2, pancreatic cancer with gemcitabine); two active, but recruiting is finished (Phase 1, pancreatic cancer with gemcitabine and radiation; Phase 1, glioblastoma multiforme, temozolomide and radiation); two active and recruiting (Phase 2, Non-Small-Cell Lung Cancer Paclitaxel, Carboplatin); Phase 2, glioblastoma multiforme, temozolomide and radiation). An overview of the current results will be presented.

Conclusions: From our basic science results, P-AscH$^-$ may be an effective adjuvant for some standard care therapies; P-AscH$^-$ is safe as an adjuvant with the chemotherapeutic agents we have tested, as well as with radiation; adverse events are minimal and mild; there are suggestions of efficacy.

3.2. Epigenetic Treatment of Cancer by Vitamin C

Gaofeng Wang, PhD *

Miller School of Medicine, University of Miami, FL 33136, USA; gwang@med.miami.edu
* Unfortunately, Dr. Wang was unable to attend the meeting due to hurricane Irma.

Recent advances have uncovered a previously unknown function of vitamin C in regulating the demethylation of DNA and histones. Ten-eleven translocation (TET) dioxygenases initiate DNA demethylation by converting 5-methylcytosine (5mC) into 5-hydroxymethylcytosine (5hmC). Vitamin C is essential for the function of TET by providing Fe(II), a cofactor of TET. Loss of 5hmC is accompanied by malignant cellular transformation. Overexpressing TET can partially re-establish a normal 5hmC profile in cancer cells and represses their malignancy. While overexpressing TETs in patients might not be clinically feasible, these discoveries suggest that finding a means of restoring normal 5hmC content may yield a novel therapy for cancer. The expression of vitamin C transporter SVCT2 is frequently downregulated in cancer. For instance, SVCT2 expression is decreased in 72.5% of breast cancer cases by at least 1.5-fold compared to the matched normal breast tissues. This suggests that it is necessary to compensate the downregulated SVCT2 with vitamin C supplements in at-risk populations and cancer patients. Treatment of cancer cells with vitamin C increases 5hmC content and results in a markedly shifted transcriptome. These changes are correlated with decreased cellular malignant phenotypes. Furthermore, by promoting the demethylation of DNA and histones, vitamin C changes the response of cancer treatment. For example, vitamin C improves the efficacy of Bromodomain and extraterminal domain inhibitors (BETi) in treating melanoma. In conclusion, vitamin C can prevent cancer initiation and progression by reestablishing 5hmC. Vitamin C also can improve the response of certain cancer drugs.

3.3. IVC and Chemotherapy in Ovarian Cancer

Jeanne Drisko, MD

University of Kansas Medical Center, Kansas City, KS 66160, USA; jdrisko@kumc.edu

Background: Ascorbate (vitamin C) has long been used as an unorthodox therapy for cancer, even though the underlying scientific mechanisms are not well understood.

Clinical Trial: A pilot phase 1/2a clinical trial was conducted in patients with newly diagnosed stage III or IV ovarian cancer. High-dose intravenous ascorbate (AA) was added to conventional paclitaxel/carboplatin (Pax + Cp) therapy, and toxicity was assessed. Twenty-seven participants were randomized into either the standard Cp + Pax arm or the Cp + Pax + AA arm. Cp + Pax chemotherapy was administered for the initial 6 months, and AA treatment for 12 months. Any and all unwanted events were counted and graded for severity according to NCI CTCAEv3. Records for adverse events include patient interviews, emergency room visits, patients' oncologist visits, and hospitalization records. The number of adverse events in each grade for each participant was divided by the number of encounters of that participant, and then the adverse events per encounter were averaged in the Cp + Pax arm and the Cp + Pax + AA arm, respectively. Participants were followed for survival for 5 years.

Statistical Analysis: Two-tailed Student's *t* test was performed for toxicity comparison between chemotherapy group and chemotherapy + ascorbate group. Welch's *t* test was used when the variances in the two compared populations were unequal. A log-rank test was performed for comparison of the survival curves between the chemotherapy group and the chemotherapy + ascorbate group.

Results: Ascorbate worked synergistically in vitro and in vivo with the first-line chemotherapeutic drugs carboplatin and paclitaxel. In patients with advanced ovarian cancer, treatment with ascorbate reduced toxicities associated with chemotherapy. Because the study was not powered for detection of efficacy, statistical improvement in survival was not observed.

Conclusion: Given the advantage of low toxicity of ascorbate, larger clinical trials need to be done to definitively examine the benefit of adding ascorbate to conventional chemotherapy.

3.4. Challenges of Natural Product Drug Development

Channing Paller, MD

Department of Oncology, Johns Hopkins University School of Medicine, Baltimore, MD 21205, USA; cpaller1@jhmi.edu

Numerous drugs that the US Food and Drug Administration (FDA) has approved for use in cancer therapy are derived from plants, including taxanes such as paclitaxel and vinca alkaloids such as vinblastine. Dietary supplements are another category of natural products that are widely used by patients with cancer, but without the FDA-reviewed evidence of safety and efficacy—be it related to survival, palliation, symptom mitigation, and/or immune system enhancement—that is required for therapy approval. Nearly half of patients in the United States with cancer report that they started taking new dietary supplements after being given a diagnosis of cancer. Oncologists are challenged in providing advice to patients about which supplements are safe and effective to use to treat cancer or the side effects of cancer therapy, and which supplements are antagonistic to standard treatment with chemotherapy, radiation, and/or immunotherapy. Despite the large number of trials that have been launched, the FDA has not approved any dietary supplements or food for the prevention of cancer, to halt its growth, or to prevent its recurrence. We will review the primary challenges faced by researchers attempting to conduct rigorous trials of natural products, including shortages of funding due to lack of patentability, manufacturing difficulties, contamination, and lack of product consistency. We will also highlight the methods used by dietary supplement marketers to persuade patients that a supplement is effective (or at least safe) even without FDA approval, as well as the efforts of the US government to protect the health and safety of its citizens by ensuring that the information used to market natural products is accurate. Finally we will close with a summary of the most widely used databases of information about the safety, efficacy, and interactions of dietary supplements.

4. Vitamin C Therapy in Sepsis

4.1. IVC in Pre-Clinical Sepsis and Trauma Models

Ramesh Natarajan, PhD

Department of Internal Medicine, Virginia Commonwealth University School of Medicine, Richmond, VA 23298-0663, USA; ramesh.natarajan@vcuhealth.org

Bacterial infections of the lungs and abdomen are among the most common causes of sepsis. Sepsis-induced acute lung injury (ALI) is a persisting clinical problem with no direct therapy. We used various models of sepsis in wild-type and knockout mice to determine whether parenteral vitamin C modulates the dysregulated pro-inflammatory, pro-coagulant state that leads to sepsis-induced lung injury. Male C57BL/6 wild type mice and mice lacking functional L-gulono-γ-lactone oxidase ($Gulo^{-/-}$) were exposed to bacterial lipopolysaccharide (endotoxin) or a fecal stem solution (polymicrobial sepsis) to induce abdominal peritonitis 30 min prior to parenterally receiving either reduced vitamin C (ascorbic acid, 200 mg/kg) or oxidized vitamin C (dehydroascorbic acid, 200 mg/kg). Outcomes examined included survival, extent of ALI, pulmonary inflammatory markers, bronchoalveolar epithelial permeability, alveolar fluid clearance, epithelial ion channel, and pump expression, tight junction protein expression, cytoskeletal rearrangements, neutrophil extracellular trap formation (NETosis), multiple organ failure and various coagulation parameters in whole blood. Sepsis-induced ALI was characterized by compromised lung epithelial permeability, reduced alveolar fluid clearance, pulmonary inflammation and neutrophil sequestration, increased formation of NETs, significant viscoelastic changes in blood, and increased mortality due to multiple organ failure. A single infusion of parenteral vitamin C protected mice from the deleterious consequences of sepsis by multiple mechanisms, including attenuation of the NFκB driven pro-inflammatory response, enhancement of epithelial barrier function, increasing alveolar fluid clearance, restoration of endothelial function, prevention of sepsis-associated coagulation abnormalities, attenuation of NETosis, and normalization of physiological functions that attenuated the development of multiple organ dysfunction. These pre-clinical studies using parenteral vitamin C were central to a completed PHASE I trial of intravenous vitamin C in sepsis patients and an ongoing PHASE II multi-center trial examining the efficacy of intravenous vitamin C infusion in human sepsis-associated ALI.

Proteomic analysis of plasma from a model of hemorrhagic shock and polytrauma in swine showed that IV vitamin C may mitigate the pro-inflammatory/pro-coagulant response that contributes to multiple organ failure following acute severe polytrauma by maintaining circulating levels of ADAMTS13. ADAMTS13 is a disintegrin-like metalloprotease that cleaves and inactivates von Willebrand Factor, a pro-coagulant molecule that is released from activated endothelial cells and promotes platelet activation.

4.2. Intravenous Vitamin C as Therapy for Sepsis-Induced Acute Lung Injury

Alpha A. (Berry) Fowler, III, MD

Department of Internal Medicine, Virginia Commonwealth University School of Medicine, Richmond, VA 23298-0663, USA; alpha.fowler@vcuhealth.org

The incidence of sepsis and sepsis-associated organ failure continues to rise in American Intensive Care Units. Over 1 million cases of sepsis (i.e., bacterial, fungal, viral) occur in the U.S. population each year. Some 375,000 patients will die from sepsis, either primarily from septic shock or secondarily from organ failure. Sepsis induced organ failure contributes cumulatively to patient mortality. Patients with severe sepsis suffer higher mortality rates compared to patients with organ failure but no sepsis. Despite over 15,000 patients studied, and over 1 billion dollars in study costs, effective sepsis therapy remains elusive. Clinical trials that have targeted mediators of inflammation or coagulation such as rosuvastatin or activated protein C have not reduced septic mortality, suggesting that single-target therapy fails to meet the challenges of complex multicellular activation and interactions. Recent studies suggest that ascorbic acid may attenuate pathological responses in septic microvasculature. In preclinical studies, ascorbic acid improved capillary blood flow, microvascular barrier function, and arteriolar responsiveness to vasoconstrictors in septic animals. We showed that parenterally administered ascorbic acid attenuated vascular lung injury in septic mice. Subnormal plasma ascorbic acid concentrations in septic patients correlates inversely with multiple organ failure and directly with survival. We report, in this presentation, that intravenous vitamin C is safe to administer to patients with severe sepsis, that it improves sepsis-induced organ failure, and that it attenuates biomarkers of systemic inflammation and vascular injury. We also report aspects of the ongoing NIH-sponsored trial: Vitamin C Infusion for TReatment In Sepsis Induced Acute Lung Injury (CITRIS-ALI).

4.3. Vitamin C Requirements and Mechanisms of Action in Severe Infection

Anitra C. Carr, PhD

Department of Pathology, University of Otago, Christchurch 8140, New Zealand; anitra.carr@otago.ac.nz

Patients with severe infections such as pneumonia can develop sepsis, an uncontrolled inflammatory response to the initial infection. This can result in organ failure and septic shock, the major cause of death of critically ill patients in intensive care. We and others have found that critically ill patients have severely depleted vitamin C levels, despite recommended intakes via liquid nutrition [4]. One study has shown that critically ill patients require parenteral vitamin C at levels that are 30-fold higher than recommended intakes. Patients with sepsis have dysregulated immune function, including compromised leukocyte function. Neutrophils are the primary responders to infection, and these cells are known to accumulate high levels of vitamin C, suggesting an important role for the vitamin in immune cell function. We and others have shown that vitamin C can enhance neutrophil chemotaxis, oxidant production, apoptosis and clearance by macrophages [5]. It is possible that the dramatic clearance of septic lungs observed following vitamin C treatment could be partly due to enhanced clearance of apoptotic neutrophils. Vitamin C is an important cofactor for numerous biosynthetic and regulatory enzymes in the body. Because vitamin C is a cofactor for the enzymes that synthesize noradrenaline and vasopressin, we hypothesized that vitamin C administration to patients with severe sepsis and septic shock may decrease the need for exogenous administration of these vasopressors [6]. Support for the vitamin C and vasopressor hypothesis comes from recent clinical trials that showed decreased vasopressor requirements in patients who received intravenous vitamin C. We are currently implementing a clinical trial in Christchurch Hospital ICU to assess the outcomes and mechanisms of action of intravenous vitamin C in severe sepsis.

4.4. Vitamin C, Hydrocortisone and Thiamine for the Treatment of Severe Sepsis and Septic Shock: The Metabolic Resuscitation Protocol

Paul E. Marik, PhD

Department of Internal Medicine, Eastern Virginia Medical School, Norfolk, VA 23501-1980, USA; marikpe@evms.edu

A large body of experimental data has demonstrated that both corticosteroids and intravenous vitamin C reduce activation of nuclear factor κB (NF-κB), attenuating the release of pro-inflammatory mediators; reduce the endothelial injury characteristic of sepsis, thereby reducing endothelial permeability and improving microcirculatory flow; augment the release of endogenous catecholamines; and enhance vasopressor responsiveness. In animal models, these effects have resulted in reduced organ injury and increased survival. Corticosteroids have been evaluated in several clinical trials, with meta-analysis of these trials demonstrating somewhat conflicting outcomes. Low-dose stress corticosteroids have proven to be safe, with no increased risk of clinically important complications. While corticosteroids decrease vasopressor dependency, the effect on the risk of developing organ failure and survival is less clear. Similarly, intravenous vitamin C has been evaluated in unselected surgical ICU patients, patients with burns, those with pancreatitis, and in two pilot studies of patients with severe sepsis and septic shock [7]. In general, these studies have demonstrated a reduction in the risk of multisystem organ failure (MSOF), although the effect on mortality is less clear. However, IV vitamin C has been shown to be extremely safe, with no recorded complications.

In vitro data has suggested that vitamin C and hydrocortisone may act synergistically. Barabutis et al. have demonstrated that hydrocortisone, together with vitamin C, protects the vascular endothelium from damage by endotoxin while neither agent alone had this effect [8]. Based on these clinical and experimental data, we initiated a treatment protocol for patients with severe sepsis and septic shock that included intravenous vitamin C, hydrocortisone and thiamine. We have demonstrated that this therapeutic cocktail reverses the organ dysfunction of sepsis with a marked reduction in mortality [9].

Conflicts of Interest: The author declares no conflict of interest.

References

1. Huijskens, M.J.; Walczak, M.; Koller, N.; Briedé, J.J.; Senden-Gijsbers, B.L.; Schnijderberg, M.C.; Bos, G.M.J.; Germeraad, W.T. Ascorbic acid induces development of double-positive T cells from human hematopoietic stem cells in the absence of stromal cells. *J. Leukoc. Biol.* **2014**, *96*, 1165–1175. [CrossRef] [PubMed]

2. Huijskens, M.J.; Walczak, M.; Sarkar, S.; Atrafi, F.; Senden-Gijsbers, B.L.; Tilanus, M.G.; Bos, G.M.J.; Wieten, L.; Germeraad, W.T. Ascorbic acid promotes the generation and proliferation of NK cell populations in different culture systems applicable for NK cell therapy. *Cytotherapy* **2015**, *17*, 613–620. [CrossRef] [PubMed]
3. Huijskens, M.J.; Wodzig, W.K.; Walczak, M.; Germeraad, W.T.; Bos, G.M.J. Ascorbic acid serum levels are reduced in patients with hematological malignancies. *Results Immunol.* **2016**, *12*, 8–10. [CrossRef] [PubMed]
4. Carr, A.C.; Rosengrave, P.C.; Bayer, S.; Chambers, S.; Mehrtens, J.; Shaw, G.M. Hypovitaminosis C and vitamin C deficiency in critically ill patients despite recommended enteral and parenteral intakes. *Crit. Care* **2017**, in press.
5. Carr, A.C.; Maggini, S. Vitamin C and immune function. *Nutrients* **2017**, *9*, 1211. [CrossRef] [PubMed]
6. Carr, A.C.; Shaw, G.M.; Fowler, A.A.; Natarajan, R. Ascorbate-dependent vasopressor synthesis: A rationale for vitamin C administration in severe sepsis and septic shock? *Crit. Care* **2015**, *19*, 418. [CrossRef] [PubMed]
7. Fowler, A.A.; Syed, A.A.; Knowlson, S.; Sculthorpe, R.; Farthing, D.; DeWilde, C.; Farthing, C.A.; Larus, T.L.; Martin, E.; Brophy, D.F.; et al. Phase 1 safety trial of intravenous ascorbic acid in patients with severe sepsis. *J. Transl. Med.* **2014**, *12*, 32. [CrossRef] [PubMed]
8. Barabutis, N.; Khangoora, V.; Marik, P.E.; Catravas, J.D. Hydrocortisone and ascorbic acid synergistically protect against LPS-induced pulmonary endothelial barrier dysfunction. *Chest* **2017**, *152*, 954–962. [PubMed]
9. Marik, P.E.; Khangoora, V.; Rivera, R.; Hooper, M.H.; Catravas, J. Hydrocortisone, vitamin C and thiamine for the treatment of severe sepsis and septic shock: A retrospective before-after study. *Chest* **2017**, *151*, 1229–1238. [CrossRef] [PubMed]

antioxidants

MDPI

Review

Vitamin C and Microvascular Dysfunction in Systemic Inflammation

Karel Tyml [1,2]

[1] Centre for Critical Illness Research, Lawson Health Research Institute, London, ON N6A 5W9, Canada;
     karel.tyml@lhsc.on.ca
[2] Department of Medical Biophysics, University of Western Ontario, London, ON N6A 5C1, Canada

Received: 6 June 2017; Accepted: 27 June 2017; Published: 29 June 2017

Abstract: Sepsis, life-threatening organ dysfunction caused by a dysfunctional host response to infection, is associated with high mortality. A promising strategy to improve the outcome is to inject patients intravenously with ascorbate (vitamin C). In animal models of sepsis, this injection improves survival and, among others, the microvascular function. This review examines our recent work addressing ascorbate's ability to inhibit arteriolar dysfunction and capillary plugging in sepsis. Arteriolar dysfunction includes impaired vasoconstriction/dilation (previously reviewed) and impaired conduction of vasoconstriction/dilation along the arteriole. We showed that ascorbate injected into septic mice prevents impaired conducted vasoconstriction by inhibiting neuronal nitric oxide synthase-derived NO, leading to restored inter-endothelial electrical coupling through connexin 37-containing gap junctions. Hypoxia/reoxygenation (confounding factor in sepsis) also impairs electrical coupling by protein kinase A (PKA)-dependent connexin 40 dephosphorylation; ascorbate restores PKA activation required for this coupling. Both effects of ascorbate could explain its ability to protect against hypotension in sepsis. Capillary plugging in sepsis involves P-selectin mediated platelet-endothelial adhesion and microthrombi formation. Early injection of ascorbate prevents capillary plugging by inhibiting platelet-endothelial adhesion and endothelial surface P-selectin expression. Ascorbate also prevents thrombin-induced platelet aggregation and platelet surface P-selectin expression, thus preventing microthrombi formation. Delayed ascorbate injection reverses capillary plugging and platelet-endothelial adhesion; it also attenuates sepsis-induced drop in platelet count in systemic blood. Thrombin-induced release of plasminogen-activator-inhibitor-1 from platelets (anti-fibrinolytic event in sepsis) is inhibited by ascorbate pH-dependently. Thus, under acidotic conditions in sepsis, ascorbate promotes dissolving of microthrombi in capillaries. We propose that protected/restored arteriolar conduction and capillary bed perfusion by ascorbate contributes to reduced organ injury and improved survival in sepsis.

Keywords: sepsis; microvessels; endothelial cells; platelets; electrical coupling; connexins; nitric oxide; P-selectin; coagulation; plasminogen-activator-inhibitor-1

---

1. Introduction

Local infectious or non-infectious insult can lead to a systemic inflammatory response. Sepsis, life-threatening organ dysfunction caused by a dysfunctional host response to infection [1], can precipitate multiple organ failure and 40% mortality in Intensive Care Units [2]. Sepsis is annually responsible for the loss of more lives than breast, colorectal, pancreatic and prostate cancers combined [3]. The prevalence of septic patients is highest in the elderly (i.e., older than 65 years) where the outcome disproportionately worsens with age [4]. Sepsis involves many pathophysiological processes including increased oxidative stress [5]. The age-aggravated worsening of outcome could be due to increased mitochondrial free radical formation that occurs naturally in aging tissues [6,7].

Among all antioxidants, ascorbate (reduced vitamin C) is considered to be the most effective water-soluble antioxidant [8]. In healthy middle-age humans, plasma ascorbate concentration is 60–80 µM [6,9], but in healthy elderly it drops to ~40 µM [10,11]. Critically, in the septic elderly, plasma ascorbate is clinically considered to be depleted (i.e., ~10 µM) [12–14]. Thus, the major defense by the antioxidant ascorbate against sepsis is nearly absent in the elderly.

A promising strategy to improve the outcome of sepsis is to replete ascorbate in patients quickly after the diagnosis of sepsis [15]. Indeed, a recent clinical study including the septic elderly reported markedly improved survival in patients injected intravenously with vitamin C, hydrocortisone and thiamine [16]. This improved survival is consistent with that observed in septic mice injected with ascorbate [15,17,18].

The sepsis-induced inflammatory response leads to dysfunction of many organ systems, including the cardiovascular system where decreased systemic vascular resistance, hypotension, maldistribution of blood flow in the microcirculation, and impaired oxygen utilization occur [19,20]. Using various animal models of sepsis, our laboratory has examined the dysfunction of the microcirculation, and the possible beneficial effects of intravenous injection of ascorbate against this dysfunction. The models, the dysfunction, and the effects of various doses of ascorbate have been reviewed [5,15,21,22]. However, our recent advances in this area extend our understanding of protection by ascorbate against microvascular dysfunction in sepsis. The objective of the present paper is to review these advances.

2. Arteriolar Dysfunction in Sepsis

Within the systemic vascular system, arterioles represent the key site along the vascular tree responsible for both the control of blood supply to the tissue and the peripheral vascular resistance [23]. The vascular resistance (and flow control) depend on (i) the degree of arteriolar diameter change elicited by local physiological/pharmacological stimuli impinging on the arteriolar wall, and (ii) the degree of conduction (or spread) of the diameter change along the arteriolar length [24]. A local arteriolar dilation without conduction yielded no increase in blood flow in the microvascular network fed by the stimulated arteriole [25].

We have shown that sepsis (cecal ligation and perforation, CLP) in young mice impairs norepinephrine-induced vasoconstriction in 6–10 µm arterioles in skeletal muscle, and that ascorbate intravenous injection protects against this impairment [26]. Similar protection by ascorbate in the vasculature has been shown for other vasoconstrictors as well as for vasodilators [21,27,28]. The mechanism of protection by ascorbate against impaired vasoconstriction has been reviewed [15]. It involves (i) inhibition of nicotinamide adenine dinucleotide phosphate (NADPH) oxidase and inducible nitric oxide synthase (iNOS) in endothelial cells of the vascular wall and (ii) inhibition of the subsequent refractory vasodilation caused by iNOS-derived NO.

2.1. Arteriolar Conducted Response in Vivo

In addition to the arteriolar diameter response, we have also examined the effects of sepsis and ascorbate on arteriolar conduction. The conduction is underpinned by electrical coupling along the arteriolar endothelial layer where connexins (i.e., constituents of inter-cellular gap junctions) are required for this coupling [29]. We used CLP (24 h model of sepsis) and lipopolysaccharide (LPS) in young mice to show that sepsis impairs conducted vasoconstriction in skeletal muscle by tyrosine kinase- and NO-dependent mechanisms [24,30]. We further determined that this impairment is mediated by the neuronal NOS (nNOS)-derived NO production and that the target of NO signaling could be the gap junction protein connexin 37 (Cx37) in the arteriolar wall [31]. Finally, we showed that an intravenous bolus of ascorbate prevented as well as reversed impairment of conducted vasoconstriction at 24 h of sepsis by inhibiting nNOS-derived NO production [32].

These studies indicate that, in addition to iNOS, nNOS is also an important source of vascular NO in our in vivo model of sepsis. Here, nNOS is found in smooth muscle cells and adjacent skeletal muscle cells [33,34]. It is possible that the protective effect of ascorbate against impaired conduction involves the heat shock protein 90 (HSP 90). HSP 90 is up-regulated during sepsis [35] and, when it binds to nNOS, it increases nNOS activity [36]. Because sepsis increases the level of reactive oxygen species (ROS) in skeletal muscle [26], and HSP 90 protein expression increases in response to ROS [37], ascorbate could scavenge ROS, prevent HSP 90 protein up-regulation, inhibit sepsis-induced increased nNOS activity and NO production, and thus prevent the septic impairment of arteriolar conduction [29].

2.2. Inter-Endothelial Electrical Coupling In Vitro

In order to gain further mechanistic insights into the impairment of conduction, we have developed an electrophysiological approach to determine inter-endothelial electrical coupling under conditions that mimic sepsis. We used monolayers of cultured microvascular endothelial cells obtained from the mouse skeletal muscle (i.e., the same tissue studied in vivo). Regarding the role of nNOS-derived NO in the impairment, we mimicked sepsis by applying exogenous NO to the monolayer. NO reduced electrical coupling [38]. Using cells from mice where individual vascular connexins were knocked-out, we determined that NO indeed targets Cx37 to reduce coupling [38]. This reduction could be due to the effect of peroxynitrite (i.e., formed after NO reaction with superoxide), or due to a direct effect of NO. Pretreatment of monolayers with ascorbate or with peroxynitrite scavenger did not affect the reduction in coupling, indicating that NO reduces coupling directly, possibly via Cx37 nitrosylation [38]. Thus, ascorbate appears to protect against impaired arteriolar conduction sepsis by affecting arteriolar function indirectly (i.e., reducing nNOS activity and NO production), rather than by affecting the target molecule Cx37.

In addition to addressing the mechanism of impaired conduction during the advanced stage in sepsis involving NO (i.e., 24 h post-CLP), we also used our electrophysiological approach in endothelial cell monolayers to address the mechanism of impaired conduction caused by LPS (i.e., an initiating factor in sepsis). We discovered that LPS reduces inter-endothelial electrical coupling via tyrosine-, ERK1/2-, PKA-, and PKC-dependent signaling that targets Cx40 [39]. This finding was consistent with the LPS-induced tyrosine-dependent impaired conduction observed in arterioles in vivo [24].

Importantly, impaired arteriolar dilatation/constriction and conduction in sepsis results in impaired microvascular blood flow which, in turn, precipitates episodes of micro-regional ischemia/reperfusion (I/R) in the tissue supplied by the arteriole (i.e., evidenced by intermittent capillary blood flow in septic skeletal muscle, [40]). I/R has been shown to aggravate the sepsis-induced inflammatory response [41,42]. Because ascorbate prevents the development of intermittent capillary blood flow in sepsis in vivo [40], we also sought to determine if ascorbate protects against reduction in inter-endothelial electrical coupling in our endothelial cell monolayer model in vitro. Using a hypoxia/reoxygenation (H/R) protocol to mimic I/R, we discovered that (i) H/R reduces inter-endothelial coupling PKA-dependently, also by targeting Cx40, and (ii) ascorbate pretreatment of the monolayer prevents this reduction by scavenging ROS [43]. This scavenging eliminates PKA inhibition by ROS [43]. Significantly, we were able to corroborate the aggravating effect by I/R on sepsis-induced inflammatory response. Concurrent LPS+H/R application to the monolayer synergistically reduced inter-endothelial electrical coupling, PKA- and PKC-dependently [44]. We demonstrated that LPS+H/R initiates tyrosine kinase- and ERK1/2-sensitive signaling that reduces electrical coupling by dephosphorylating PKA-specific serine residues of Cx40 [44]. Our most recent work pinpointed the residues 345–358 of the Cx40 carboxyl terminal tail as possible sites of this dephosphorylation [45].

Taken together, a complex picture emerges for the impaired arteriolar conduction in sepsis. Initially, LPS and the concurrent H/R may reduce inter-endothelial electrical coupling and arteriolar conduction by targeting Cx40, whereas nNOS-mediated NO overproduction in advanced sepsis reduces coupling and conduction by targeting Cx37 instead. The protection by ascorbate against impaired conduction involves (i) inhibition of the H/R component in the initial stage of sepsis (i.e., ascorbate restores PKA activation required for conduction) and, in the advanced stage, (ii) inhibition of nNOS activation and excess NO production.

Protection by ascorbate against both sepsis-induced impairment in arteriolar vasoconstriction and conduction could explain ascorbate's ability (i) to inhibit hypotension in rat models of sepsis [40,46] and (ii) to markedly reduce duration of vasopressor treatment in septic patients, normally necessitated by falling blood pressure [16].

3. Capillary Plugging in Sepsis

Sepsis-induced inflammation leads to activation of the coagulation pathway [47]. We have used the skeletal muscle in rats and mice as a bioassay to examine this aspect of sepsis in terms of capillary bed plugging, a well-known indicator of sepsis involving pro-coagulant responses [5,15]. Capillary plugging involves P-selectin mediated platelet-endothelial adhesion and fibrin deposition in capillaries [48,49]. This plugging, reported in animal and human organs, leads to inadequate oxygenation of the tissue and organ failure [5,15]. Importantly, we and others have shown that intravenous injection of ascorbate protects against sepsis-induced capillary plugging and organ injury [17,40,48,50]. The experimental details and the mechanism of this protection against capillary plugging have been reviewed [5,15]. A key component of this mechanism is endothelial nitric oxide synthase (eNOS) in endothelial cells of the microvascular wall. Since eNOS-derived NO is anti-coagulatory (i.e., it reduces platelet-endothelial adhesion [51]), and since the protection by ascorbate is absent in eNOS$^{-/-}$ mice [50], ascorbate has been proposed to act indirectly via restoring the eNOS function in the microvasculature in sepsis [50].

Ascorbate intravenous injection early in sepsis prevents capillary plugging, whereas delayed ascorbate injection later in sepsis reverses plugging (i.e., restores blood flow in previously plugged capillaries) [40,46,48]. Recently, we have addressed the mechanisms of both the prevention and reversal of plugging.

3.1. Ascorbate Prevents Capillary Plugging in Sepsis

A key event in the initiation process of sepsis-induced capillary plugging is platelet adhesion to the capillary wall. Pretreatment of mice with platelet-depleting antibody inhibits this plugging [48]. P-selectin is a key platelet-endothelium adhesion molecule [52]. Pretreatment of mice with P-selectin blocking antibody also inhibits plugging [48]. We have carried out a series of experiments designed to tease out whether platelet surface and/or endothelial surface P-selectin expression are involved in the initiation. Using a platelet-endothelial cell adhesion assay in vitro, we determined that activation of endothelial cells by LPS increased platelet adhesion to the endothelial monolayer (mouse skeletal muscle origin) P-selectin-dependently [49]. Further, LPS increased P-selectin protein expression at the surface of endothelial cells, most likely by promoting exocytosis of P-selectin protein already contained in Weibel-Palade granules beneath the endothelial surface [49,53,54]. Significantly, pretreatment of the monolayer with ascorbate inhibited all platelet adhesion to the monolayer, endothelial P-selectin surface expression, and exocytosis [49]. These in vitro studies demonstrated that ascorbate can inhibit platelet-endothelial adhesion in capillaries in vivo directly, rather than indirectly via hemodynamic effects of ascorbate on capillary blood flow.

We have also used a platelet aggregation assay in vitro, to examine the role of P-selectin at the platelet surface. LPS or plasma from septic mice did not alter P-selectin expression at the platelet surface, or platelet aggregation [55]. However, platelet-activating agents known to be released into the bloodstream during sepsis [thrombin, adenosine diphosphate (ADP), thromboxane A2] did increase P-selectin expression and aggregation. Interestingly, ascorbate inhibited these increases independently of platelet-derived NOS [55]. Thus, ascorbate could reduce aggregation directly, independent of its ability to restore eNOS function within the microvasculature. In the context of plugging of septic capillaries, the inhibition of platelet aggregation directly by ascorbate may not be enough to fully prevent plugging. In addition to this direct effect, NO-derived from non-platelet sources (e.g., endothelial eNOS) may be needed for the full in vivo effect of ascorbate.

Our results suggest a complex mechanism in the initiation of capillary plugging and in the protection by ascorbate against this plugging. An early platelet-endothelial adhesion may be followed/paralleled by the generation of platelet-activating stimuli which, in turn, result in platelet aggregation and buildup of other materials (e.g., fibrin deposits, microthrombi) which eventually plug capillaries [48,55]. So far our data indicate that ascorbate can protect against the initial platelet-endothelial adhesion and the subsequent platelet aggregation in the septic capillary. These data are consistent with the reported anti-coagulatory effect of ascorbate in sepsis [17,18].

3.2. Ascorbate Reverses Capillary Plugging

A delayed intravenous injection of ascorbate reverses both the number of plugged capillaries and platelet adhesion/trapping therein (observed in skeletal muscle, in a mouse model of sepsis involving feces injection into peritoneum, FIP) [48]. This reversal is eNOS-dependent [48]. Since platelet trapping in capillaries leads to subsequent reduction in the number of platelets available for detection in systemic blood, the platelet count in systemic blood is a complementary, clinically relevant [56], measure of capillary plugging. To this end, we showed that (i) sepsis at 7 h post-FIP indeed reduces platelet count measured in samples of arterial blood and (ii) ascorbate injection delayed to 6 h post-FIP attenuates this reduction (i.e., previously trapped platelets were released back into the systemic circulation) [48,57].

To address ascorbate's ability to quickly reverse platelet trapping (i.e., over 1 h period) we examined ascorbate's ability to dissolve microthrombi in capillaries. This dissolving will permit restarting of blood flow in these microvessels. To this end, we examined the thrombolytic system. Using our FIP model of sepsis in mice, sepsis increased mRNA of both the pro-fibrinolytic urokinase plasminogen activator (u-PA) and the anti-fibrinolytic plasminogen activator inhibitor 1 (PAI-1) in muscle and liver homogenates [57]. Delayed ascorbate did not affect u-PA mRNA in either tissue; it inhibited PAI-1 mRNA in the muscle (i.e., suggesting enhanced fibrinolysis in this tissue) but not in the liver. Since liver PAI-1 is the dominant source of soluble PAI-1 in systemic blood, we further examined PAI-1 enzymatic activity in this tissue. Ascorbate did not affect sepsis-induced increase in PAI-1 activity in the liver [57]. Consistently, delayed ascorbate also did not affect sepsis-induced increase in PAI-1 protein and activity in systemic blood plasma [57,58]. Thus, based on the PAI-1 protein/enzymatic activity data measured in tissue homogenates and systemic blood, our study did not support the hypothesis that ascorbate reverses capillary plugging in sepsis by promoting fibrinolysis.

Local pro- and anti-fibrinolytic events, which occur at the level of the capillary, may not necessarily be assessed by analyzing tissue homogenates or systemic blood. A clear example of this is the observation that sepsis causes hypocoagulability in systemic blood but hypercoagulability in the microcirculation [15]. To address this issue, we used our in vitro models of cultured microvascular endothelial cells (mouse skeletal muscle origin) and platelets isolated from mice (i.e., both cell types are present in the milieu of capillary microthrombi) [59]. Because both cell types can release PAI-1 into the extracellular space [60,61], we asked whether ascorbate affects PAI-1 release from these cells. We used thrombin or LPS to mimic sepsis. In unstimulated endothelial cells and platelets, PAI-1 was released into the extracellular space and this release was unaffected by ascorbate pretreatment. Thrombin or LPS did not alter PAI-1 release from endothelial cells. However, thrombin, but not LPS, increased PAI-1 release from platelets. Ascorbate inhibited this release pH-dependently [59].

Thus, under acidotic conditions prevalent in sepsis, our in vitro studies suggest that, together with the inhibition by ascorbate of thrombin-induced platelet aggregation discussed above, inhibition by ascorbate of thrombin-induced PAI-1 release from platelets would yield a pro-fibrinolytic effect leading to dissolving of microthrombi in septic capillaries.

3.3. A Multifaceted Mechanism of Capillary Plugging

Our work has focused mainly on the role of endothelial cells and platelets in capillary plugging observed by intravital microscopy in the septic skeletal muscle. Clearly, there are other cell types which could contribute to the formation of microvascular microthrombi. These include red blood cells which become stiff in sepsis and thus may obstruct the capillary lumen [15,62]. Additionally, activated leukocytes, including neutrophils, can adhere to the capillary/venular endothelium and thus increase the hemodynamic resistance to blood flow. The number of adhering leukocytes was negligible in capillaries and venules in the skeletal muscle of septic mice; these cells thus could not account for the capillary plugging, or be involved in the inhibitory effect of ascorbate against plugging in this tissue [48,62]. However, the presence of neutrophils in immunogenic organs such as the lung and liver and their abundance there during sepsis [18,59] would undoubtedly contribute to capillary plugging therein. Recent reports of neutrophil extracellular traps (NETs) contributing to platelet aggregation or leukocyte-platelet aggregation [63], and to microthrombi formation [64], underscore the involvement of neutrophils in capillary plugging. Importantly, ascorbate has been shown to reduce the lung NETs formation in septic mice [65].

Sepsis leads to endothelial barrier dysfunction, involving increased permeability in the microvasculature and increased extravasation of plasma proteins and fluid (reviewed by [15]). This extravasation could form tissue edema, compress the capillary lumen, and thus also contribute to capillary plugging in sepsis. Ascorbate can inhibit this dysfunction by inhibiting NADPH oxidase expression and activity in endothelial cells, by attenuating protein phosphatase 2 (PP2A) activation, and subsequently restoring the phosphorylation and distribution of the tight junction protein occludin in endothelial cells (mechanism reviewed by [15]).

4. Unresolved Issues and Future Directions in Experimental Studies of Systemic Inflammation

Most experimental studies of sepsis have used young animals, but the majority of septic patients are elderly. In septic mice, it has been shown that the levels of plasma inflammatory cytokines, antioxidant defense, and mortality markedly worsen in aged when compared to young mice [66–68]. Thus, studies in young animals may have a limited impact on our understanding and development of therapeutic strategies to treat sepsis in the elderly.

To our knowledge, there are no reports addressing the effect of ascorbate on the outcome of sepsis in aged mice. Relevant to this unresolved issue is a recent study using Gulo$^{-/-}$ mice [69]. These mice are deficient in endogenous vitamin C production and thus require the vitamin supplementation in diet. In Gulo$^{-/-}$ mice without supplementation (i.e., mimicking the nearly-depleted plasma ascorbate status in the septic elderly), sepsis resulted in exacerbated mortality and organ injury when compared to both Gulo$^{-/-}$ mice with ascorbate supplementation and Gulo$^{-/-}$ mice injected with ascorbate after the onset of sepsis [69]. The study suggests that ascorbate repletion would be critical when treating the septic elderly.

Because of the increased incidence of obesity in the present general population, another unresolved issue may be the effect of ascorbate on the outcome of sepsis in obese animals or in animals with other co-morbidities. Similar to the effect of aging, obesity in mice also worsens the inflammatory response to sepsis [70]. However, in a clinical study, sepsis in obese patients [71] did not worsen the outcome as predicted by this animal study, underscoring the complexity of human sepsis. To our knowledge, there are no reports addressing the effect of ascorbate on the outcome of sepsis in obese animals.

The non-infectious insult I/R also leads to a systemic inflammatory response, including lung injury [72], impaired arteriolar conduction [43] and capillary plugging [73]. Ascorbate has been shown to attenuate the I/R-induced injury [72] and H/R-induced impairment of inter-endothelial electrical coupling [43]. A key feature of the H/R-induced coupling impairment is H/R-stimulated increase in ROS production in endothelial cells [43,74]. Intriguingly, the stimulated ROS increase is Cx40-dependent, possibly involving a cross-talk between Cx40 and NADPH oxidase [74]. Thus, Cx40 may not function only as a structural protein in intercellular gap junctions [29,75], but also as a signaling molecule responsible for the H/R-stimulated ROS increase in endothelial cells. Given the reported aggravation by I/R in sepsis-induced inflammatory response [41,42], and the critical roles of NAPDH oxidase in impaired arteriolar vasoconstriction in sepsis and of Cx40 in impaired electrical coupling in endothelial cells exposed to LPS, H/R or LPS+H/R, the possible signaling function of Cx40 warrants further investigation.

5. Conclusions

Despite numerous animal studies and clinical trials, the mortality in sepsis remains unacceptably high. A promising strategy to improve the outcome of sepsis is to intravenously inject patients with ascorbate (vitamin C) to quickly restore its levels in blood plasma and tissues. We have shown that intravenous injection of ascorbate improves the microvascular function in septic rats and mice. These improvements include the arteriolar responsiveness to vasoactive stimuli and capillary bed perfusion.

In particular, our recent work demonstrated that ascorbate inhibits the sepsis-induced impairment of arteriolar conducted vasoconstriction by inhibiting nNOS-derived NO production and ROS production, to restore the inter-endothelial cell electrical coupling and gap junction function. These effects contribute to ascorbate's ability to protect against hypotension in sepsis. Further, we demonstrated that ascorbate inhibits capillary plugging in sepsis by inhibiting platelet-endothelial adhesion and platelet aggregation mediated by P-selectin, and by promoting the dissolution of microthrombi in capillaries. These effects contribute to ascorbate's ability to protect against tissue injury and to improve survival in sepsis.

Acknowledgments: I would like to thank Scott Swarbreck, Dan Secor, Darcy Lidington, Michael Bolon, Feng Wu, Fuyan Li, Rebecca McKinnon, Gail Yu, Mohammad Siddiqui, John Armour, and Nigel Gocan for their lab work and accomplishments summarized in this review, John Wilson, Christopher Ellis, Sean Gill, Yves Ouellette, Gerald Kidder and Dale Laird for stimulating discussions, and the Heart and Stroke Foundation of Ontario and the Canadian Institutes of Health Research for providing financial support. I also acknowledge William Sibbald who, in the 1990s, assembled a group of clinicians and basic scientists to spearhead long-term research into the mechanism of circulatory dysfunction in sepsis, including the microcirculatory dysfunction, at Victoria Hospital Research Institute, London, Ontario.

Conflicts of Interest: The author declares no conflict of interest.

References

1. Singer, M.; Deutschman, C.S.; Seymour, C.W.; Shankar-Hari, M.; Annane, D.; Bauer, M.; Bellomo, R.; Bernard, G.R.; Chiche, J.D.; Coopersmith, C.M.; et al. The third international consensus definitions for sepsis and septic shock (sepsis-3). *JAMA* **2016**, *315*, 801–810. [CrossRef] [PubMed]
2. Martin, C.M.; Priestap, F.; Fisher, H.; Fowler, R.A.; Heyland, D.K.; Keenan, S.P.; Longo, C.J.; Morrison, T.; Bentley, D.; Antman, N. A prospective, observational registry of patients with severe sepsis: The Canadian Sepsis Treatment and Response Registry. *Crit. Care Med.* **2009**, *37*, 81–88. [CrossRef] [PubMed]
3. Angus, D.C.; Linde-Zwirble, W.T.; Lidicker, J.; Clermont, G.; Carcillo, J.; Pinsky, M.R. Epidemiology of severe sepsis in the United States: Analysis of incidence, outcome, and associated costs of care. *Crit. Care Med.* **2001**, *29*, 1303–1310. [CrossRef] [PubMed]
4. Martin, G.S.; Mannino, D.M.; Moss, M. The effect of age on the development and outcome of adult sepsis. *Crit. Care Med.* **2006**, *34*, 15–21. [CrossRef] [PubMed]
5. Tyml, K. Critical role for oxidative stress, platelets, and coagulation in capillary blood flow impairment in sepsis. *Microcirculation* **2011**, *18*, 152–162. [CrossRef] [PubMed]
6. Bailey, D.M.; McEneny, J.; Mathieu-Costello, O.; Henry, R.R.; James, P.E.; McCord, J.M.; Pietri, S.; Young, I.S.; Richardson, R.S. Sedentary aging increases resting and exercise-induced intramuscular free radical formation. *J. Appl. Physiol.* **2010**, *109*, 449–456. [CrossRef] [PubMed]
7. Miquel, J.; Economos, A.C.; Fleming, J.; Johnson, J.E. Mitochondrial role in cell aging. *Exp. Gerontol.* **1980**, *15*, 575–591. [CrossRef]

8. Frei, B.; England, L.; Ames, B.N. Ascorbate is an outstanding antioxidant in human blood plasma. *Proc. Natl. Acad. Sci. USA* **1989**, *86*, 6377–6381. [CrossRef] [PubMed]
9. Levine, M.; Conry-Cantilena, C.; Wang, Y.; Welch, R.W.; Washko, P.W.; Dhariwal, K.R.; Park, J.B.; Lazarev, A.; Graumlich, J.F.; King, J.; et al. Vitamin C pharmacokinetics in healthy volunteers: Evidence for a recommended dietary allowance. *Proc. Natl. Acad. Sci. USA* **1996**, *93*, 3704–3709. [CrossRef] [PubMed]
10. Heseker, H.; Schneider, R. Requirement and supply of vitamin C, E and beta-carotene for elderly men and women. *Eur. J. Clin. Nutr.* **1994**, *48*, 118–127. [PubMed]
11. Smith, V.H. Vitamin C deficiency is an under-diagnosed contributor to degenerative disc disease in the elderly. *Med. Hypotheses* **2010**, *74*, 695–697. [CrossRef] [PubMed]
12. Fain, O.; Pariés, J.; Jacquart, B.T.; Le Moël, G.; Kettaneh, A.; Stirnemann, J.; Héron, C.; Sitbon, M.; Taleb, C.; Letellier, E.; et al. Hypovitaminosis C in hospitalized patients. *Eur. J. Intern. Med.* **2003**, *14*, 419–425. [CrossRef] [PubMed]
13. Galley, H.F.; Davies, M.J.; Webster, N.R. Ascorbyl radical formation in patients with sepsis: Effect of ascorbate loading. *Free Radic. Biol. Med.* **1996**, *20*, 139–143. [CrossRef]
14. Paz, H.L.; Martin, A.A. Sepsis in an aging population. *Crit. Care Med.* **2006**, *34*, 234–235. [CrossRef] [PubMed]
15. Wilson, J.X. Evaluation of vitamin C for adjuvant sepsis therapy. *ARS* **2013**, *19*, 2129–2140. [CrossRef] [PubMed]
16. Marik, P.E.; Khangoora, V.; Rivera, R.; Hooper, M.H.; Catravas, J. Hydrocortisone, Vitamin C and thiamine for the treatment of severe sepsis and septic shock: A retrospective before–after study. *Chest* **2016**, *151*, 1229–1238. [CrossRef] [PubMed]
17. Fisher, B.J.; Seropian, I.M.; Kraskauskas, D.; Thakkar, J.N.; Voelkel, N.F.; Fowler, A.A.; Natarajan, R. Ascorbic acid attenuates lipopolysaccharide-induced acute lung injury. *Crit. Care Med.* **2011**, *39*, 1454–1460. [CrossRef] [PubMed]
18. Fisher, B.J.; Kraskauskas, D.; Martin, E.J.; Farkas, D.; Wegelin, J.A.; Brophy, D.; Ward, K.R.; Voelkel, N.F.; Fowler, A.A.; Natarajan, R. Mechanisms of attenuation of abdominal sepsis induced acute lung injury by ascorbic acid. *Am. J. Physiol. Lung Cell. Mol. Physiol.* **2012**, *303*, 20–32. [CrossRef] [PubMed]
19. Nguyen, H.B.; Rivers, E.P.; Knoblich, B.P.; Jacobsen, G.; Muzzin, A.; Ressler, J.A.; Tomlanovich, M.C. Early lactate clearance is associated with improved outcome in severe sepsis and septic shock. *Crit. Care Med.* **2004**, *32*, 1637–1642. [CrossRef] [PubMed]
20. Bone, R.C. Gram-negative sepsis: Background, clinical features, and intervention. *Chest* **1991**, *100*, 802–808. [CrossRef] [PubMed]
21. Wilson, J.X. Mechanism of action of vitamin C in sepsis: Ascorbate modulates redox signaling in endothelium. *Biofactors* **2009**, *35*, 5–13. [CrossRef] [PubMed]
22. Wilson, J.X.; Wu, F. Vitamin C in sepsis. *Subcell. Biochem.* **2012**, *56*, 67–83. [PubMed]
23. Joyner, W.L.; Davis, M.J. Pressure profile along the microvascular network and its control. *Fed. Proc.* **1987**, *46*, 266–269. [PubMed]
24. Tyml, K.; Wang, X.; Lidington, D.; Ouellette, Y. Lipopolysaccharide reduces intercellular coupling *in vitro* and arteriolar conducted response in vivo. *Am. J. Physiol. Heart Circ. Physiol.* **2001**, *281*, H1397–H1406. [PubMed]
25. Kurjiaka, D.T.; Segal, S.S. Conducted vasodilation elevates flow in arteriole networks of hamster striated muscle. *Am. J. Physiol.* **1995**, *269*, H1723–H1728. [PubMed]
26. Wu, F.; Wilson, J.X.; Tyml, K. Ascorbate inhibits iNOS expression and preserves vasoconstrictor responsiveness in skeletal muscle of septic mice. *Am. J. Physiol. Regul. Integr. Comp. Physiol.* **2003**, *285*, R50–R56. [CrossRef] [PubMed]
27. Wu, F.; Wilson, J.X.; Tyml, K. Ascorbate protects against impaired arteriolar constriction in sepsis by inhibiting inducible nitric oxide synthase expression. *Free Radic. Biol. Med.* **2004**, *37*, 1282–1289. [CrossRef] [PubMed]
28. Aschauer, S.; Gouya, G.; Klickovic, U.; Storka, A.; Weisshaar, S.; Vollbracht, C.; Krick, B.; Weiss, G.; Wolzt, M. Effect of systemic high dose vitamin C therapy on forearm blood flow reactivity during endotoxemia in healthy human subjects. *Vascul. Pharmacol.* **2014**, *61*, 25–29. [CrossRef] [PubMed]
29. Tyml, K. Role of connexins in microvascular dysfunction during inflammation. *Can. J. Physiol. Pharmacol.* **2011**, *89*, 1–12. [CrossRef] [PubMed]
30. Lidington, D.; Ouellette, Y.; Li, F.; Tyml, K. Conducted vasoconstriction is reduced in a mouse model of sepsis. *J. Vasc. Res.* **2003**, *40*, 149–158. [CrossRef] [PubMed]

31. McKinnon, R.L.; Lidington, D.; Bolon, M.; Ouellette, Y.; Kidder, G.M.; Tyml, K. Reduced arteriolar conducted vasoconstriction in septic mouse cremaster muscle is mediated by nNOS-derived NO. *Cardiovasc. Res.* **2006**, *69*, 236–244. [CrossRef] [PubMed]

32. McKinnon, R.L.; Lidington, D.; Tyml, K. Ascorbate inhibits reduced arteriolar conducted vasoconstriction in septic mouse cremaster muscle. *Microcirculation* **2007**, *14*, 697–707. [CrossRef] [PubMed]

33. Gocan, N.C.; Scott, J.A.; Tyml, K. Nitric oxide produced via neuronal NOS may impair vasodilatation in septic rat skeletal muscle. *Am. J. Physiol. Heart Circ. Physiol.* **2000**, *278*, H1480–H1489. [PubMed]

34. Kavdia, M.; Popel, A.S. Contribution of nNOS- and eNOS-derived NO to microvascular smooth muscle NO exposure. *J. Appl. Physiol.* **2004**, *97*, 293–301. [CrossRef] [PubMed]

35. Hashiguchi, N.; Ogura, H.; Tanaka, H.; Koh, T.; Nakamori, Y.; Noborio, M.; Shiozaki, T.; Nishino, M.; Kuwagata, Y.; Shimazu, T.; et al. Enhanced expression of heat shock proteins in activated polymorphonuclear leukocytes in patients with sepsis. *J. Trauma* **2001**, *51*, 1104–1109. [CrossRef] [PubMed]

36. Song, Y.; Zweier, J.L.; Xia, Y. Heat-shock protein 90 augments neuronal nitric oxide synthase activity by enhancing Ca^{2+}/calmodulin binding. *Biochem. J.* **2001**, *355*, 357–360. [CrossRef] [PubMed]

37. Muller, M.; Gauley, J.; Heikkila, J.J. Hydrogen peroxide induces heat shock protein and proto-oncogene mRNA accumulation in Xenopus laevis A6 kidney epithelial cells. *Can. J. Physiol. Pharmacol.* **2004**, *82*, 523–529. [CrossRef] [PubMed]

38. McKinnon, R.L.; Bolon, M.L.; Wang, H.-X.; Swarbreck, S.; Kidder, G.M.; Simon, A.M.; Tyml, K. Reduction of electrical coupling between microvascular endothelial cells by NO depends on connexin37. *Am. J. Physiol. Heart Circ. Physiol.* **2009**, *297*, H93–H101. [CrossRef] [PubMed]

39. Bolon, M.L.; Kidder, G.M.; Simon, A.M.; Tyml, K. Lipopolysaccharide reduces electrical coupling in microvascular endothelial cells by targeting connexin40 in a tyrosine-, ERK1/2-, PKA-, and PKC-dependent manner. *J. Cell. Physiol.* **2007**, *211*, 159–166. [CrossRef] [PubMed]

40. Armour, J.; Tyml, K.; Lidington, D.; Wilson, J.X. Ascorbate prevents microvascular dysfunction in the skeletal muscle of the septic rat. *J. Appl. Physiol.* **2001**, *90*, 795–803. [PubMed]

41. Khadaroo, R.G.; Kapus, A.; Powers, K.A.; Cybulsky, M.I.; Marshall, J.C.; Rotstein, O.D. Oxidative stress reprograms lipopolysaccharide signaling via Src kinase-dependent pathway in RAW 264.7 macrophage cell line. *J. Biol. Chem.* **2003**, *278*, 47834–47841. [CrossRef] [PubMed]

42. Powers, K.A.; Szaszi, K.; Khadaroo, R.G.; Tawadros, P.S.; Marshall, J.C.; Kapus, A.; Rotstein, O.D. Oxidative stress generated by hemorrhagic shock recruits Toll-like receptor 4 to the plasma membrane in macrophages. *JEM* **2006**, *203*, 1951–1961. [CrossRef] [PubMed]

43. Bolon, M.L.; Ouellette, Y.; Li, F.; Tyml, K. Abrupt reoxygenation following hypoxia reduces electrical coupling between endothelial cells of wild-type but not connexin40 null mice in oxidant- and PKA-dependent manner. *FASEB J.* **2005**, *19*, 1725–1727. [CrossRef] [PubMed]

44. Bolon, M.L.; Peng, T.; Kidder, G.M.; Tyml, K. Lipopolysaccharide plus hypoxia and reoxygenation synergistically reduce electrical coupling between microvascular endothelial cells by dephosphorylating connexin40. *J. Cell. Physiol.* **2008**, *217*, 350–359. [CrossRef] [PubMed]

45. Siddiqui, M.; Swarbreck, S.; Shao, Q.; Secor, D.; Peng, T.; Laird, D.W.; Tyml, K. Critical role of Cx40 in reduced endothelial electrical coupling by lipopolysaccharide and hypoxia-reoxygenation. *J. Vasc. Res.* **2015**, *52*, 396–403. [CrossRef] [PubMed]

46. Tyml, K.; Li, F.; Wilson, J.X. Delayed ascorbate bolus protects against maldistribution of microvascular blood flow in septic rat skeletal muscle. *Crit. Care Med.* **2005**, *33*, 1823–1828. [CrossRef] [PubMed]

47. Levi, M.; van der Poll, T.; Büller, H.R. Bidirectional relation between inflammation and coagulation. *Circulation* **2004**, *109*, 2698–2704. [CrossRef] [PubMed]

48. Secor, D.; Li, F.; Ellis, C.G.; Sharpe, M.D.; Gross, P.L.; Wilson, J.X.; Tyml, K. Impaired microvascular perfusion in sepsis requires activated coagulation and P-selectin-mediated platelet adhesion in capillaries. *Intensive Care Med.* **2010**, *36*, 1928–1934. [CrossRef] [PubMed]

49. Secor, D.; Swarbreck, S.; Ellis, C.G.; Sharpe, M.D.; Feng, Q.; Tyml, K. Ascorbate inhibits platelet-endothelial adhesion in an in-vitro model of sepsis via reduced endothelial surface P-selectin expression. *Blood Coagul. Fibrinolysis* **2017**, *28*, 28–33. [CrossRef] [PubMed]

50. Tyml, K.; Li, F.; Wilson, J.X. Septic impairment of capillary blood flow requires nicotinamide adenine dinucleotide phosphate oxidase but not nitric oxide synthase and is rapidly reversed by ascorbate through an endothelial nitric oxide synthase-dependent mechanism. *Crit. Care Med.* **2008**, *36*, 2355–2362. [CrossRef] [PubMed]

51. Cerwinka, W.H.; Cooper, D.; Krieglstein, C.F.; Feelisch, M.; Granger, D.N. Nitric oxide modulates endotoxin-induced platelet-endothelial cell adhesion in intestinal venules. *Am. J. Physiol. Heart Circ. Physiol.* **2002**, *282*, H1111–H1117. [CrossRef] [PubMed]

52. Blann, A.D.; Nadar, S.K.; Lip, G.Y.H. The adhesion molecule P-selectin and cardiovascular disease. *Eur. Heart J.* **2003**, *24*, 2166–2179. [CrossRef] [PubMed]

53. McCarron, R.M.; Doron, D.A.; Sirén, A.L.; Feuerstein, G.; Heldman, E.; Pollard, H.B.; Spatz, M.; Hallenbeck, J.M. Agonist-stimulated release of von willebrand factor and procoagulant factor VIII in rats with and without risk factors for stroke. *Brain Res.* **1994**, *647*, 265–272. [CrossRef]

54. Wang, G.F.; Wu, S.Y.; Rao, J.J.; Lü, L.; Xu, W.; Pang, J.X.; Liu, Z.Q.; Wu, S.G.; Zhang, J.J. Genipin inhibits endothelial exocytosis via nitric oxide in cultured human umbilical vein endothelial cells. *Acta. Pharmacol. Sin.* **2009**, *30*, 589–596. [CrossRef] [PubMed]

55. Secor, D.; Swarbreck, S.; Ellis, C.G.; Sharpe, M.D.; Tyml, K. Ascorbate reduces mouse platelet aggregation and surface P-selectin expression in an ex vivo model of sepsis. *Microcirculation* **2013**, *20*, 502–510. [CrossRef] [PubMed]

56. Moreau, D.; Timsit, J.F.; Vesin, A.; Garrouste-Orgeas, M.; de Lassence, A.; Zahar, J.R.; Adrie, C.; Vincent, F.; Cohen, Y.; Schlemmer, B.; et al. Platelet count decline: an early prognostic marker in critically ill patients with prolonged ICU stays. *Chest* **2007**, *131*, 1735–1741. [CrossRef] [PubMed]

57. Swarbreck, S.; Secor, D.; Li, F.; Gross, P.L.; Ellis, C.G.; Sharpe, M.D.; Wilson, J.X.; Tyml, K. Effect of ascorbate on fibrinolytic factors in septic mouse skeletal muscle. *Blood Coagul. Fibrinolysis* **2014**, *25*, 745–753. [CrossRef] [PubMed]

58. Swarbreck, S.B.; Secor, D.; Ellis, C.G.; Sharpe, M.D.; Wilson, J.X.; Tyml, K. Short-term effect of ascorbate on bacterial content, plasminogen activator inhibitor-1, and myeloperoxidase in septic mice. *J. Surg. Res.* **2014**, *191*, 432–440. [CrossRef] [PubMed]

59. Swarbreck, S.B.; Secor, D.; Ellis, C.G.; Sharpe, M.D.; Wilson, J.X.; Tyml, K. Effect of ascorbate on plasminogen activator inhibitor-1 expression and release from platelets and endothelial cells in an in-vitro model of sepsis. *Blood Coagul. Fibrinolysis* **2015**, *26*, 436–442. [CrossRef] [PubMed]

60. Sagripanti, A.; Morganti, M.; Carpi, A.; Cupisti, A.; Nicolini, A.; Barsotti, M.; Camici, M.; Mittermayer, C.; Barsotti, G. Uremic medium increases cytokine-induced PAI-1 secretion by cultured endothelial cells. *Biomed. Pharmacother.* **1998**, *52*, 298–302. [CrossRef]

61. Nylander, M.; Osman, A.; Ramström, S.; Aklint, E.; Larsson, A.; Lindahl, T.L. The role of thrombin receptors PAR1 and PAR4 for PAI-1 storage, synthesis and secretion by human platelets. *Thromb. Res.* **2012**, *129*, 51–58. [CrossRef] [PubMed]

62. Bateman, R.M.; Jagger, J.E.; Sharpe, M.D.; Ellsworth, M.L.; Mehta, S.; Ellis, C.G. Erythrocyte deformability is a nitric oxide-mediated factor in decreased capillary density during sepsis. *Am. J. Physiol. Heart Circ. Physiol.* **2001**, *280*, H2848–H2856. [PubMed]

63. Tanaka, K.; Koike, Y.; Shimura, T.; Okigami, M.; Ide, S.; Toiyama, Y.; Okugawa, Y.; Inoue, Y.; Araki, T.; Uchida, K.; et al. In vivo characterization of neutrophil extracellular traps in various organs of a murine sepsis model. *PLoS ONE* **2014**. [CrossRef] [PubMed]

64. McDonald, B.; Davis, R.P.; Kim, S.J.; Tse, M.; Esmon, C.T.; Kolaczkowska, E.; Jenne, C.N. Platelets and neutrophil extracellular traps collaborate to promote intravascular coagulation during sepsis in mice. *Blood* **2017**, *129*, 1357–1367. [CrossRef] [PubMed]

65. Mohammed, B.M.; Fisher, B.J.; Kraskauskas, D.; Farkas, D.; Brophy, D.F.; Fowler, A.A.; Natarajan, R. Vitamin C: A novel regulator of neutrophil extracellular trap formation. *Nutrients* **2013**, *5*, 3131–3151. [CrossRef] [PubMed]

66. Starr, M.E.; Ueda, J.; Takahashi, H.; Weiler, H.; Esmon, C.T.; Evers, B.M.; Saito, H. Age-dependent vulnerability to endotoxemia is associated with reduction of anticoagulant factors activated protein C and thrombomodulin. *Blood* **2010**, *115*, 4886–4893. [CrossRef] [PubMed]

67. Starr, M.E.; Ueda, J.; Yamamoto, S.; Evers, B.M.; Saito, H. The effects of aging on pulmonary oxidative damage, protein nitration, and extracellular superoxide dismutase down-regulation during systemic inflammation. *Free Radic. Biol. Med.* **2011**, *50*, 371–380. [CrossRef] [PubMed]

68. Turnbull, I.R.; Clark, A.T.; Stromberg, P.E.; Dixon, D.J.; Woolsey, C.A.; Davis, C.G.; Hotchkiss, R.S.; Buchman, T.G.; Coopersmith, C.M. Effects of aging on the immunopathologic response to sepsis. *Crit. Care Med.* **2009**, *37*, 1018–1023. [CrossRef] [PubMed]

69. Fisher, B.J.; Kraskauskas, D.; Martin, E.J.; Farkas, D.; Puri, P.; Massey, H.D.; Idowu, M.O.; Brophy, D.F.; Voelkel, N.F.; Fowler, A.A.; et al. Attenuation of Sepsis-induced Organ Injury in Mice by Vitamin C. *JPEN* **2013**, *38*, 825–839. [CrossRef] [PubMed]

70. Vachharajani, V.; Russell, J.M.; Scott, K.L.; Conrad, S.; Stokes, K.Y.; Tallam, L.; Hall, J.; Granger, D.N. Obesity exacerbates sepsis-induced inflammation and microvascular dysfunction in mouse brain. *Microcirculation* **2005**, *12*, 183–194. [CrossRef] [PubMed]

71. Arabi, Y.M.; Dara, S.I.; Tamim, H.M.; Rishu, A.H.; Bouchama, A.; Khedr, M.K.; Feinstein, D.; Parrillo, J.E.; Wood, K.E.; Keenan, S.P.; et al. Clinical characteristics, sepsis interventions and outcomes in the obese patients with septic shock: An international multicenter cohort study. *Crit Care.* **2013**, *17*, R72. [CrossRef] [PubMed]

72. Baltalarli, A.; Ozcan, V.; Bir, F.; Ferda, B.; Aybek, H.; Sacar, M.; Onem, G.; Goksin, I.; Demir, S.; Teke, Z.; et al. Ascorbic acid (vitamin C) and iloprost attenuate the lung injury caused by ischemia/reperfusion of the lower extremities of rats. *Ann. Vasc. Surg.* **2006**, *20*, 49–55. [CrossRef] [PubMed]

73. Bihari, A.; Cepinskas, G.; Forbes, T.L.; Potter, R.F.; Lawendy, A.R. Systemic application of carbon monoxide-releasing molecule 3 protects skeletal muscle from ischemia-reperfusion injury. *J. Vasc. Surg.* **2017**. [CrossRef] [PubMed]

74. Yu, G.; Bolon, M.; Laird, D.W.; Tyml, K. Hypoxia and reoxygenation-induced oxidant production increase in microvascular endothelial cells depends on connexin40. *Free Radic. Biol. Med.* **2010**, *49*, 1008–1013. [CrossRef] [PubMed]

75. Guo, R.; Si, R.; Scott, B.T.; Makino, A. Mitochondrial connexin40 regulates mitochondrial calcium uptake in coronary endothelial cells. *Am. J. Physiol. Cell. Physiol.* **2017**, *312*, C398–C406. [CrossRef] [PubMed]

antioxidants

MDPI

Review

Influence of Vitamin C on Lymphocytes: An Overview

Gwendolyn N. Y. van Gorkom [1,*], Roel G. J. Klein Wolterink [1], Catharina H. M. J. Van Elssen [1], Lotte Wieten [2], Wilfred T. V. Germeraad [1] and Gerard M. J. Bos [1]

[1] Division of Hematology, Department of Internal Medicine,
GROW-School for Oncology and Developmental Biology, Maastricht University Medical Center,
6202AZ Maastricht, The Netherlands; r.kleinwolterink@maastrichtuniversity.nl (R.G.J.K.W.);
janine.van.elssen@mumc.nl (C.H.M.J.V.E.); w.germeraad@maastrichtuniversity.nl (W.T.V.G.);
gerard.bos@mumc.nl (G.M.J.B.)
[2] Department of Transplantation Immunology, Maastricht University Medical Center, 6202 AZ Maastricht,
The Netherlands; Lotte.wieten@maastrichtuniversity.nl
* Correspondence: gwendolyn.van.gorkom@mumc.nl; Tel.: +31-043-387-6543

Received: 8 February 2018; Accepted: 8 March 2018; Published: 10 March 2018

Abstract: Vitamin C or ascorbic acid (AA) is implicated in many biological processes and has been proposed as a supplement for various conditions, including cancer. In this review, we discuss the effects of AA on the development and function of lymphocytes. This is important in the light of cancer treatment, as the immune system needs to regenerate following chemotherapy or stem cell transplantation, while cancer patients are often AA-deficient. We focus on lymphocytes, as these white blood cells are the slowest to restore, rendering patients susceptible to often lethal infections. T lymphocytes mediate cellular immunity and have been most extensively studied in the context of AA biology. In vitro studies demonstrate that T cell development requires AA, while AA also enhances T cell proliferation and may influence T cell function. There are limited and opposing data on the effects of AA on B lymphocytes that mediate humoral immunity. However, AA enhances the proliferation of NK cells, a group of cytotoxic innate lymphocytes. The influence of AA on natural killer (NK) cell function is less clear. In summary, an increasing body of evidence indicates that AA positively influences lymphocyte development and function. Since AA is a safe and cheap nutritional supplement, it is worthwhile to further explore its potential benefits for immune reconstitution of cancer patients treated with immunotoxic drugs.

Keywords: vitamin C; ascorbic acid; lymphocytes; natural killer cells; NK cells; B cells; T cells

1. Introduction

Vitamin C or ascorbic acid (AA) has often been linked to cancer treatment. Already in the 1970s, Cameron and Pauling reported that high doses of AA intravenously increased the survival time of terminal cancer patients more than four times [1] but this finding could not be repeated in other studies where AA supplementation was given orally [2,3]. However, subsequent studies show that AA has a wide variety of effects on cancer cells and the immune system. In this review, we discuss the effects of AA on lymphocytes in the light of cancer treatment.

AA is an essential micronutrient for humans with many functions in the human body. It is an antioxidant and a free radical scavenger and serves as an essential cofactor for many enzymatic reactions through iron-, copper- and 2-oxoglutarate-dependent dioxygenases. Among many other functions, these dioxygenases are important in epigenetic regulation by catalysing the hydroxylation of methylated nucleic acids (DNA and RNA) and histones [4]. While most mammals use the enzyme gulono-gamma-lactone oxidase to synthesize AA in the liver, many primates and humans carry a non-functional copy of the *GULO*-gene and consequently depend on dietary sources of AA. When studying the effects of AA and AA deficiency in vivo in animal models, this is a complicating factor. Guinea pigs, like humans, also have a defect in the GULO-gene and are thereby often chosen for AA deficiency studies. Alternatively, there are two knockout mouse models, a *Gulo* knockout ($Gulo^-/^-$) and a senescence marker protein-30 knockout (*SMP30KO*), in which biosynthesis of AA in the liver is blocked [5,6].

AA has an extensive role in the immune system. Its role in phagocytic cells like neutrophils, has been investigated thoroughly and was recently reviewed [7]. In summary, AA enhances chemotaxis and phagocytosis of phagocytes and thereby promotes microbial killing. In contrast, the role of AA in different subsets of lymphocytes is less clear. Since lymphocytes actively acquire AA via sodium-dependent vitamin C transporters (SVCT) and sodium independent glucose transporters (GLUT) (reviewed in [8] and have intracellular AA concentrations that are 10–100-fold higher than plasma levels [9,10], it is likely that AA has an essential function in these cells. There are three main subsets of lymphocytes, namely T cells, B cells and natural killer (NK cells). T cells are involved in cell-mediated, cytotoxic adaptive immunity, B cells are responsible for the adaptive, humoral immunity and NK cells are part of the innate, antigen-independent immunity.

In our laboratory, we are interested in lymphocytes in cancer treatment, because these cells are often destroyed by anticancer treatment and take time to recover. During this phase, patients are highly susceptible to possibly lethal infections. Depending on the intensity of the chemotherapy used, this period may be relatively short, for example in breast cancer, or long, for example in leukaemia. After so-called myeloablative chemotherapy, hematopoietic stem cells (HSC) that are located in the bone marrow have to be replaced in order to restore all types of blood cells, including leukocytes. In particular, the regeneration of T-lymphocytes, a subset of lymphocytes that are especially important to fight against viral infections, is slow as a consequence of age-dependent involution of the thymus, the organ that is required for their development [11,12]. Looking at ways to improve T cell recovery after cancer therapy, we investigated factors that influence human T-lymphocyte development and found that AA acts as a factor that promotes maturation of T cells. AA is also indispensable for T cell development in vitro [13]. Additionally, we showed that NK cells regenerate faster under the influence of AA [14].

We also found that haematological cancer patients often have severely decreased serum AA levels compared to healthy controls (20.5 ± 12 µM versus 65 ± 4 µM, respectively). Serum AA levels were even undetectable in 19% of patients with a haematological malignancy, irrespective of the choice of treatment [15]. Since AA is a cheap and readily available supplement with a safe profile, it is attractive to speculate that cancer patients who need to regenerate their immune system after chemotherapy with or without hematopoietic stem cell transplantation (HSCT) may benefit from the effects of AA on immune reconstitution. In this way, we hypothesize that mortality and morbidity resulting from opportunistic infections could be reduced. It could also be that NK cells regenerate faster and are able to kill cancer cells sooner. AA supplementation could also be used in cellular therapies, where in vitro proliferated and adapted subtypes of lymphocytes are used to eliminate tumour cells in vivo. However, before using AA in clinical applications, it is important to have a better understanding of the role of AA in these lymphocytes.

In this article, we highlight the effects of AA on different subsets of lymphocytes as far as they are known for this moment. We will focus on the effects on the physiology of these cells and on the role of AA on lymphocytes in health and disease and not on the potential mechanisms behind these effects, since this was extensively reviewed before [4].

2. AA and T Lymphocytes

T lymphocytes are a major component of the human immune system and are involved in cell-mediated, cytotoxic adaptive immunity. On their surface, T cells express the T cell receptor (TCR) that is responsible for recognizing and binding specific antigens bound to major histocompatibility complex (MHC) molecules. There are different types of T lymphocytes, including cytotoxic T cells, T helper cells, memory T cells and regulatory T cells. Cytotoxic T cells are characterized by a MHC class I binding CD8 protein on their cell surface. The TCR and CD8 receptor bind infected cells and tumour cells. After binding, the cytotoxic T cells mature and, upon activation by an infected cell, secrete perforin and granzymes, that kill the infected cells. T helper cells are CD4 positive cells that regulate immune responses. Their TCR binds to MHC class II on antigen presenting cells (APCs). After binding, T-helper cells secrete cytokines that activate other immune cells, including cytotoxic T cells. Memory T cells are long-living cells that recognize previously encountered pathogens and provide lifelong immunity. Regulatory T cells shut down T cell mediated immunity toward the end of an immune reaction and help to maintain tolerance to self-antigens.

Here we describe the effects on AA on general T cell development and summarize what is known about the influence of AA on these most important subsets of T cells. We will not discuss cytotoxic T cells since we found no studies examining the effects of AA on this specific subset.

2.1. T Cell Development and AA

T cell development is a tightly controlled process that takes place in the thymus, which can be simulated in vitro using foetal thymic organ cultures [16], stromal cells [17] or in feeder-free conditions [13]. While mature T lymphocytes express either CD4 or CD8 for helper and cytotoxic subsets respectively, immature T cells are called "double negative" (DN) because they lack CD4 and CD8 expression. Traveling through the highly-organized thymus, the developing T cells undergo numerous rounds of proliferation. The thymic stromal cells provide the structural support and cytokines necessary for selection of a functional TCR that does not recognize self-antigens. This process of "education" is required to generate a diverse repertoire of TCRs to ensure immunity against a wide variety of antigens. The various stages of human T cell development are characterized by sequential acquisition of CD7, CD5, intracellular CD3, CD1a, CD4 and CD8, TCR$\alpha\beta$ and surface CD3 [18].

In search for factors that enhance T cell differentiation after stem cell transplantation, we discovered that AA enhances human T cell proliferation in vitro [13]. Beside this effect on T cell proliferation, we also found multiple effects on early T cell development. Most importantly, we showed that AA is required in vitro for the transition of DN precursors to the next, so-called "double positive" (DP, CD4$^+$ CD8$^+$) stage in feeder-free cultures as well as in cultures with stromal cells when culturing T cells from cord blood or G-CSF stimulated hematopoietic stem cells. Furthermore, we found that in a feeder-free system, early maturation of T cells after 3 weeks was improved under the influence of AA in a dose dependent way with an optimum at 95 µM [13]. These results are in line with a murine study in which the investigators cultured adult bone marrow-derived hematopoietic progenitor cells on stromal cells and showed that these cells only differentiate to the DP stage in the presence of AA. To determine the effect of AA in vivo, foetal liver chimeric mice were generated by transfer of *Slc23a2*-deficient HCS into recipient mice. In the absence of *Slc23a2*, hematopoietic cells are unable to concentrate AA. Consequently, in animals with a *Slc23a2*-deficient hematopoietic system, T cell maturation was virtually absent compared to control mice [19].

Since AA functions as an antioxidant, we tested if other antioxidants could restore T cell development. As this was not the case, the effect of AA on developing human T cells cannot be attributed to its antioxidant properties [11]. This finding is supported by Manning et al. [19] who showed that induction and maintenance of *Cd8a* gene expression is dependent on AA-dependent removal of repressive histone modifications, rather than on its function as an antioxidant.

In summary, in humans and mice, AA is required in vitro for the early development of T cells as it overcomes a development block from DN to DP. Furthermore, AA speeds up the maturation process of T lymphocytes. In mice, at least part of this effect is due to AA-dependent epigenetic regulation.

2.2. T Cell Proliferation and AA

Multiple researchers studied the effect of vitamin C on the proliferation and survival of T cells, in vitro as well as in vivo.

One study describes the effect of AA on in vitro culture of in vivo activated mouse T cells. While more than 70% apoptotic cells were found in cultures without AA, the addition of AA (450 µM) decreased apoptosis by one-third and induced more proliferation was seen compared to cultures without AA [20]. In another study, evaluating the effects of AA on murine T cells during in vitro activation, it was found that that low concentrations (62.5 µM and 125 µM) of AA do not change proliferation or viability of T cells, while higher concentrations (250 µM and 500 µM) do decrease both [21]. In a third study, researchers examined how AA prevents oxidative damage using purified human T cells. They report similar effects: medium-high concentrations of AA (57–142 µM) decrease T cell proliferation, while at higher concentrations (284 µM), AA decreases cell viability and IL-2 secretion more than 90% [22]. In another study studying the expression of SVCT on T cells, the investigators show a similar effect. Peripheral blood T cells of healthy human volunteers were activated in vitro in the absence or presence of different doses AA, before and after activation. AA did not have any effect on proliferation or apoptosis in low doses (62.5–250 µM). At high doses (500–1000 µM), the proliferation was inhibited and there was an increase in apoptosis when AA was added before activation [23].

In a study on the effect of AA-deficiency on lymphocyte numbers in guinea pigs, the investigators found that in animals with an 4-week AA-free diet, the number of T-lymphocytes decreased continuously while T cell number slightly increased in AA-supplemented animals (25 and 250 mg intraperitoneally/day) [24]. Plasma and tissue concentrations of AA were significantly lower in animals without AA compared to AA-treated animals. In another in vivo study using AA-deficient $SMP30KO^{-/-}$ mice, the researchers determined the long-term effect of AA on immune cells using a diet with an increased AA level (200 mg/kg vs. 20 mg/kg). During the one-year study, T-lymphocytes in the peripheral blood increased in number. More specifically, the number of naive T cells, memory T cells in the spleen and mature T cells in the thymus [6] increased. Plasma concentrations of AA in mice with a low-dose AA diet were similar to wildtype mice, while plasma concentrations in the high-dose diet were significantly higher.

Badr et al., examined if the impaired T cell function in type I diabetes can be improved by AA supplementation using a streptozotocin-induced diabetes type I rat model. These animals have diminished T cell cytokine production, less proliferation and lower surface expression of CD28, a protein that is important for T cell activation and survival. AA supplementation (100 mg/kg/day for 2 months) restored the CD28 expression, cytokine secretion and proliferation [25].

Studies in humans are limited. Because elderly people often have lower serum levels of AA and are more prone to infections, a placebo-controlled trial was performed in which elderly people received either an intramuscular injection of AA (500 mg/day) or placebo for 1 month. Compared to the placebo group, an increase in T cell proliferation was seen in the AA-supplemented group [26]. The only other study in humans could not recapitulate this effect [27]. Healthy volunteers were kept on an AA-free diet during a 5-week period to induce AA deficiency. This did not lead to any changes in T cell numbers or function, while the induction of AA deficiency was confirmed in plasma and leukocytes.

In summary, both animal and human studies show that physiological AA concentrations have a beneficial effect on T cell proliferation, while supraphysiological concentrations are toxic for T cells. In vivo, restoration of AA in deficient patients positively influences T cell proliferation as well, while this observation could not be reproduced in induced AA-deficiency.

2.3. T Helper Cells (Th) and AA

There are several subsets of Th cells, the most important ones being Th1, Th2 and Th17. Th1 cells are part of the defence against intracellular bacteria and protozoa. Using their main effector cytokines IFN-γ and TNF-α, they activate cytotoxic T cells and macrophages. Th2 cells are effective against extracellular parasites and produce mainly IL-4, IL-5 and IL-13. They stimulate eosinophils, basophils, mast cells and B-cells. Th2 cells are important mediators of allergy and hypersensitivity. For this reason, Th2 cells are often investigated in animal models for asthma. Th17 cells have an important role in pathogen clearance of mucosal surfaces and produce IL-17, a cytokine that stimulates B-cells. The various Th subsets differentiate from naïve CD4+ T cells in a process called "polarization". In vivo, dendritic cells (DC) are the most important antigen-presenting cells (APC) that steer Th polarization via the production of cytokines.

Several researchers report that AA induces a shift of immune responses from Th2 to Th1. In one of these studies, a mouse model was used to examine the effect of AA (5 mg/day) on delayed-type hypersensitivity response against 2,4,-dinitro-I-fluorobenzene (DNFB). In this study, mice were intraperitoneally injected with AA before, during or after sensitization with DNFB. If T cells of mice supplemented with AA during the sensitization were later stimulated ex vivo, higher levels of Th1 cytokines (TNF-α and IFN-γ) and lower levels of Th2 cytokines (IL-4) were observed. This effect was not observed when mice were supplemented with AA before or after sensitization [28]. This modulation of immune balance from Th2 to Th1 was also seen in another study, in which the effects of AA supplementation on asthma was studied. Here, AA supplementation (130 mg/kg/day for 5 weeks) of ovalbumin-sensitized mice significantly increased the IFN-γ/IL-5 secretion ratio in bronchoalveolar lavage fluid compared to control mice, confirming a shift from Th2 to Th1 [29]. The mechanism underpinning this effect has not been elucidated yet but it was suggested to be mediated by DCs. In an in vitro study, murine bone marrow-derived DCs were pre-treated with different doses of AA before being activated with lipopolysaccharide (LPS). The DCs that were treated with AA secreted more IL-12, a polarizing cytokine for Th1 cells. It also showed that naïve murine T cells, when co-cultured with these activated and AA treated murine DCs, produced more IFN-γ and less IL-5 verifying this effect [30].

While most studies focus on the Th1/Th2 balance, we found only one study that describes that Th17 polarization of sorted murine naïve CD4+ cells is more effective in the presence of AA [31]. Interestingly, the investigators demonstrate that this effect is probably due to AA-mediated effects on histone demethylation that enhances the expression of the IL-17 locus.

In summary, multiple animal studies show that AA promotes Th1 differentiation at the expense of Th2 polarization. There is limited data showing that Th17 polarization is promoted by AA acting as an epigenetic regulator.

2.4. Memory T Cells and AA

Memory T cells constitute a small subset of lymphocytes but provide life-long immunity to previously encountered antigens. At this moment, the effects of AA on memory T cells is hardly investigated. Jeong et al. examined the effect of DCs pre-treated with AA on CD8+ T cell differentiation. In vitro, murine bone marrow-derived DCs were pre-treated with AA before activation with LPS and, like in the earlier study from the same research group [30], secreted more IL12p70 but also more IL-15, a cytokine that is linked to memory T cell generation. These DCs in co-culture with murine T cells led to an enhanced CD8+ memory T cell production. The effect was also seen in a mouse model for melanoma, in which the immune response and anti-melanoma effect of melanoma-primed DCs was enhanced if pre-treated with AA: the investigators observed increased generation of tumour-specific CD8+ memory T cells and an increased protective effect for inoculated melanoma cells [32].

In summary, in vitro and in a mouse model AA increased the generation of CD8+ memory T cells through increased production of stimulating cytokines by DCs.

2.5. Regulatory T Cells (Tregs) and AA

Tregs are important for the maintenance of immune-balance and self-tolerance. They are characterized by the expression of the transcription factor Foxp3, required for their immunosuppressive capacity. Stable expression of Foxp3 is dependent on DNA demethylation of a region in the first intron. Two recent in vivo and in vitro studies on murine Tregs found that AA stabilizes expression of

Foxp3 by promoting active Ten Eleven Translocation (TET) 2-mediated DNA methylation of this region. Thus, AA is required for the development and function of Tregs [33,34]. Concordantly, AA is a known co-factor for Ten Eleven Translocation (TET) family proteins that catalyse the first step of DNA demethylation: the conversion of 5-methyl-cytosine (5 mC) to 5-hydroxy-methyl-cytosine (5 hmC) [35,36]. For instance, in embryonic stem cells, AA was shown to be an important epigenetic regulator through this pathway [37]. The addition of AA to embryonic stem cell cultures induced demethylation of over 2000 genes within one hour [38].

In another study, the influence of AA on skin graft rejection in mice after treatment with ex vivo cultured and alloantigen-induced Tregs was investigated. The in vivo alloantigen-induced murine Tregs showed more DNA demethylation and stability of Foxp3 expression when cultured in the presence of AA. These Tregs also showed better suppressive capacity in vivo, thereby promoting skin allograft acceptance [39]. In a mouse model for graft-versus-host-disease (GVHD), the effects of AA on these ex vivo alloantigen-induced Tregs was determined [40]. GVHD is a serious and sometimes lethal complication following allogeneic hematopoietic stem cell transplantation caused by alloreactive donor T cells that induce tissue injury in the recipient. In this model, in vitro murine alloantigen-induced Tregs pre-treated with AA showed more stable Foxp3 expression when transferred into mice with acute GVHD and were clinically effective to diminish GVHD symptoms. Moreover, cultured human alloantigen-induced Tregs also had a higher Foxp3 expression if cultured with AA.

In contrast, in a mouse model for sepsis, AA was found to decrease the inhibition of Tregs [41]. Here, sepsis was induced in AA-deficient $Gulo^-/^-$ mice that were supplemented with AA (200 mg/kg twice) or not. AA administration improved survival in both wild type and $Gulo^-/^-$ mice and diminished the negative inhibition of Tregs by decreasing the expression of Foxp-3, CTLA-4, a protein that functions as an immune checkpoint and downregulates immune responses and the inhibitory cytokine TGF-β.

In summary, AA directly regulates Treg function via epigenetic regulation of the master transcription factor Foxp3. In most studies employing more chronic situations (transplantation, GVHD), Foxp3 expression is increased. In this way, AA can be useful in generating ex vivo allo-antigen-induced Tregs that can be used for clinical applications in transplantation and autoimmune disorders. In one model of acute sepsis in mice, AA administration decreased Foxp3 expression but was still beneficial for the outcome.

3. AA and B Lymphocytes

B lymphocytes are at the centre of the adaptive, humoral immune system. They are responsible for the production of antigen-specific immunoglobulin (Ig) directed against invasive pathogens (antibodies). Similar to other leukocytes, AA accumulated in B lymphocytes but there is only limited data on the function of AA in these cells.

3.1. B Lymphocyte Numbers and AA

In an early study investigating the effect of AA deficiency on numbers of lymphocytes in guinea pigs, animals on a 4-week AA-free diet showed a continuous increase in the percentage of B-lymphocytes while the percentage of T-lymphocytes decreased. The opposite effect was seen in animals on AA supplementation (25 and 250 mg intraperitoneally/day) [24]. In a more recent and extensive study, the effect of AA-2G, a stable vitamin C derivate, was investigated on mouse B cells in vitro. Mouse spleen B cells were cultured for 2 days with an anti-μ antibody in the presence of stimulating cytokines and then washed and recultured with and without AA-2G. In these cultures, the number of viable cells decreased much quicker without AA, resulting in about 70% more viable cells in cultures with AA than without AA. AA-2G also increased the production of IgM dose-dependently [42]. Another group that studied the effect of AA on mouse spleen B cells in vitro published contradictive results. They showed a slight dose-dependent increase of apoptosis (16% at a concentration of 1 mM) in murine IgM/CD40-activated B cells pre-treated with AA [43]. Tanaka et al. investigated the effect of AA on immune responses in human peripheral blood lymphocytes cultured for 7 days with and without AA-2G before stimulating them with pokeweed mitogen (PWM), a T cell dependent B cell stimulus. The cultures treated with AAS-2G showed an increased number of IgM and IgG-secreting cells after stimulation [44].

3.2. B Lymphocyte Function and AA

We only found limited and conflicting data on the effect of vitamin C on the production of antibodies by B lymphocytes. Two early studies in guinea pigs show that high-dose AA supplementation increases immunoglobulin levels after immunization with sheep red blood cells (SRBC) and bovine serum albumin (BSA) [45,46]. However, in two other animal models, AA supplementation did not have an effect on antigen-induced immunoglobulin levels after immunization [47,48]. One study determined the effect of high dose AA (2500 mg/day for 4 weeks) in mice sensitized by topical application of di-nitro-chlorobenzene (DNCB) and re-challenged 2 weeks later. They found no effect of AA on immunoglobulin levels [47]. In the other study, the effect of different low doses of AA (0, 30 and 60 mg) on the immune functions of dogs immunized with PWM was studied but no differences in PWM-specific IgG and IgA levels were observed [48]. However, the latter two studies were performed in animals that are able to synthetize AA, while guinea pigs cannot. In AA-synthetizing calves, AA supplementation (1.75 g/day) led to lower plasma IgG levels against keyhole limpet haemocyanin (KLH), a T cell dependent antigen that is often used in immunological studies in animals [49]. Two studies were performed in AA-sufficient chickens to investigate if AA supplementation is beneficial for vaccination against infectious bursal disease (IBD). Chickens with AA supplementation (1 g/mL) showed higher immunoglobulin levels compared to chickens without extra AA [50,51]. Furthermore, non-vaccinated chickens receiving AA supplementation did not show any symptoms or mortality after challenge with IBD while non-vaccinated chickens without AA all experienced clinical symptoms and only 70% survived [50].

In one study in healthy human volunteers, researchers found a correlation between serum IgG and plasma and leukocyte AA concentration and serum IgM and leukocyte AA concentration. After daily supplementation of 1 g AA during one week, serum IgG significantly rose in those healthy volunteers as did plasma and leukocyte AA concentration [52]. Likewise, another study in healthy human volunteers showed an increase in serum levels of IgM and IgA after 1 g/day supplementation of AA for 75 days [53]. These findings were contradicted in two other studies. In one study, the investigators examined the effect of 2–3 g AA supplementation per day on the production of all immunoglobins in healthy volunteers and did not find any change [54]. The other study is an earlier described placebo-controlled trial in elderly people where they received either an intramuscular injection of AA (500 mg/day) or a placebo for 1 month. Also in this trial, AA supplementation did not have any influence on serum IgA, IgM and IgG levels [26].

In conclusion, it is possible that vitamin C has an effect on the proliferation and function of B lymphocytes but the results until now are inconclusive. When conducting an intervention study of AA in cancer patients it would be interesting to also examine B cell levels and immunoglobulin changes.

4. AA and Natural Killer Cells

Next to B and T lymphocytes, NK cells are the most prominent lymphocyte subset as they make up to 20% of the blood lymphocyte population and are important for the immunity against pathogens (especially viruses) and for tumour surveillance. They are large granular lymphocytes arising from the same lymphoid progenitors as T and B lymphocytes and are primarily formed in the bone marrow. NK cells are innate lymphoid cells (ILCs) that provide fast, antigen-independent immunity. They exhibit direct cytotoxic effects, secrete cytokines and chemokines and regulate other immune cells. NK cell cytotoxicity is based on the absence of self MHC class I to discriminate between normal and diseased cells. Killing of these MHC class I missing target cells can only be initiated after simultaneous detection of activating signals, like stress signals, on the surface of tumour or infected cells.

Our knowledge about the effects of AA on NK cell development is limited. We previously described that AA (95 μM) enhances proliferation of mature NK cells from peripheral blood mononuclear cells (PBMCs) in vitro in a cytokine-stimulated culture [14]. AA also improved the generation and expansion of NK cell progenitors from hematopoietic stem cells and from T/NK cell progenitors in vitro in a cytokine-stimulated culture.

Other studies investigated the role of AA on the function of NK cells. In our previously described study, we tested functionality of the mature NK cells that were expanded in vitro in a cytotoxicity assay on K562 cells, a chronic myeloid leukaemia cell line that is often used to assess NK cell function in vitro. There was no difference in killing capacity between NK cells that were cultured with or without AA. [14]. In contrast, an earlier study on the cytotoxicity of fresh human NK cells isolated from peripheral blood that had different doses of AA (10 μM to 2.5 mM) present during the killing showed a dose-dependent decrease of NK cell mediated killing of K562 cells in vitro [55]. A similar experiment was repeated in another study but with different results. A presence of 3 mM AA increased the cytotoxicity of NK cells 105% in average, while no change was found using lower concentrations (10 μM, 0.1 mM and 1 mM) [56].

Another group also investigated the effect of AA on peripheral blood NK cell function using a cytotoxicity assay on K562 with NK cells from healthy volunteers that were supplemented with a single high dose of vitamin C. The study showed a biphasic effect on NK cell cytotoxicity: a slight decrease 1 to 2 h after supplementation followed by a significant enhancement at 8 h with a maximum effect after 24 h and return to normal after 48 h [57]. Plasma and leukocyte concentration of AA increased after 1 h and maximized after 2 to 4 h. After that, the levels declined but were still elevated up to 24 h after supplementation. Since NK cell function is often decreased after exposure to toxic chemicals, these researchers also performed a similar experiment in 55 patients that where accidently exposed to various toxic chemicals (for instance pesticides, metals and organic solvents). Almost half of these patients showed a very low baseline NK cell activity and in 78% of the patients there was significantly enhancement of cytotoxicity compared to baseline 24 h after ingestion of 60 mg/kg AA [58]. It is difficult to interpret these findings, since the patients were very diverse and there was no control group.

A comparable study was performed with NK cells isolated from peripheral blood of patients with β-thalassemia major. NK cells of these patients show a severe reduction in their cytotoxic function compared to healthy controls, possibly due to oxidative stress caused by iron overload after multiple blood transfusions [59]. AA (200 μg/mL) almost normalized the cytotoxic capacity of these NK cells, while the NK cell function of healthy controls did not change [60]. Remarkably, AA and iron are connected in many biological processes. For example, AA enhances the absorption of nonheme iron from the intestines [61] and AA is an essential cofactor in many enzymatic reaction by iron-dependent dioxygenases. The positive effect of AA on the NK cell function in this case is probably related to its antioxidant properties, since NK cells in patients with iron overload are known to have more intracellular reactive oxygen species (ROS) [62].

The effects of vitamin C on NK cell cytotoxicity against ovarian cancer cells was also studied in vivo in AA deficient mice. $Gulo^{-}/^{-}$ mice that are dependent on dietary AA were not supplemented for 2 weeks and sub sequentially inoculated with MOSECs (murine ovarian surface epithelial cells) and compared with $Gulo^{-}/^{-}$ mice that received normal AA supplementation. After the inoculation of tumour cells, all animals received normal AA supplementation. $Gulo^{-}/^{-}$ mice that were AA-depleted during this period survived shorter than supplemented mice. NK cells isolated from these AA-depleted mice showed a significant decrease in killing capacity in vitro compared to AA-supplemented mice and wildtype mice. In concordance, their NK cells showed reduced expression of the activating receptors CD69 and NKG2D. Furthermore, these NK cells produced less IFN-γ and displayed reduced production of the cytolytic proteins perforin and granzyme B [63].

In conclusion, AA might have different effects on NK cells during different stages. Early and late human NK cell development is enhanced by AA in vitro, however the effect in vivo remains to be shown. It is likely that the effect of AA at this time point is caused by its role as an epigenetic regulator, since this is also observed in T cells that share various developmental steps.

The influence of AA on NK cell function is not fully determined yet. In most in vitro studies in which NK cells of healthy volunteers (that probably have normal AA levels) were used, no effect of AA was observed. However, in AA-deficient mice, NK cell function was decreased compared to mice with normal levels of AA. Furthermore, in two human studies employing NK cells with an impaired function AA was able to restore NK cell function to almost normal. These results suggest that at least physiological levels of AA are necessary for normal NK cell function and AA is probably not able to increase the cytotoxicity of NK cells that function normally but can help to restore the function in NK cells that are impaired.

It is unknown whether next to NK cells, the development and function of recently identified other members of the ILC family are also enhanced by AA. This may be important, because other ILC subsets may also provide immunity while T cell immunity has not yet recovered. Furthermore, it has been shown that, in allogeneic HSCT higher ILC 3 numbers are associated with less GVHD [64].

5. Conclusions

AA has multiple effects on the development, proliferation and function of lymphocytes. An overview of these effects and the relationship between different cell types can be seen in Figure 1. T-lymphocytes have been most extensively studied in this context: AA positively influences T cell development and maturation, especially in case of AA deficiency. There is very limited and conflicting data on the effects of AA on B-lymphocyte biology. As for NK cells, AA positively influences NK cell proliferation but its role in NK cell function is less clear. A limited number of studies suggest that NK cell function required normal AA levels, while supraphysiological levels do not enhance NK cell function. Overall, most conclusions are based on in vitro studies, that are difficult to interpret and compare since there are many differences in experimental setups (multiple derivates of AA used in various concentrations and different incubation times). AA is also known to oxidate easily to dehydroascorbate in cell-cultures. The in vivo studies require careful interpretation as well: little data is available on local (intracellular) AA levels, while it is known that intracellular AA levels can be more than 1000-fold higher compared to plasma levels.

The studies discussed in the present review provide some insight in the mechanisms that underpin the effect on AA on lymphocytes. There are important indications that AA acts as an epigenetic regulator/cofactor in TET-mediated DNA and histone demethylation. Plausibly, AA's epigenetic functions are mostly seen in cells that undergo change (e.g., early T cell development, T helper cell differentiation). In situations of cells under stress (e.g., thalassemia, sepsis), the antioxidant properties of AA are probably more important.

Since AA is a cheap supplement with limited side-effects, it is worthwhile to speculate on its potential for cancer patients that are proven to have lower serum AA levels. Here, AA could enhance immune reconstitution after treatments that give long immunosuppression (for instance, patients receiving intensive chemotherapy for leukaemia or autologous HSCT) as most studies indicate a positive effect of AA supplementation on lymphocyte development. In this case, AA's effect on NK cells may be most significant, because NK cell reconstitution after myeloablative chemotherapy and HSCT is much faster than T cell reconstitution and could provide temporary immunity against infections [65]. Furthermore, NK cells are capable of recognizing and eliminating cancer cells. Currently, several clinical trials already study the anti-cancer potential of ex vivo generated NK cells. AA supplementation can be used to generate these NK cells in vitro but could also have an effect in vivo in the proliferation and survival of these cells. Furthermore, AA supplementation could may also positively influence T cell reconstitution after myeloablative therapy. It is possible that slow T cell regeneration is (partly) due to the AA-deficient state in these patients.

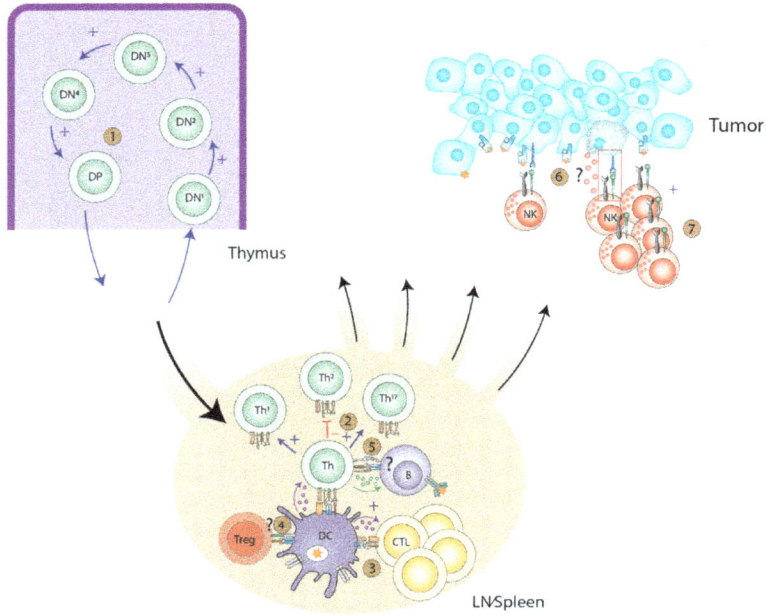

Figure 1. Effects of AA on Immune cells. **1**. T cell development: Enhanced T cell development due to fast transition from DN to DD stage. **2**. Th cell differentiation: skewing towards Th1 and Th17, with inhibition of Th2 polarization. **3**. CTL induction: Increased induction of CTLs due to production of IL-15 and IL-12 by DCs. **4**. Treg induction: Current data are conflicting. **5**. B cell: No conclusive data. **6**. NK cell function: No conclusive data. **7**. NK cell proliferation: increased NK cell proliferation.

On the other hand, AA supplementation in cancer patients may also have negative effects. Most of the curative effect of an allogeneic hematopoietic stem cell transplantation is attributed to the graft versus tumour effect mediated by the donor T cells that recognize cancer cells. This effect could potentially be diminished by increasing the amount and function of Tregs by vitamin C, as seen in some studies [33,34]. On the other hand, another study shows decreased Treg activity following AA administration [41]. In addition, stimulation of Tregs may also protect against GVHD.

Thus, AA plays a multitude of roles in lymphocyte development and function. However, its exact mechanism(s) of action and its effects in human health and disease are currently unknown. Given its safe profile and the fact that most animal studies have important limitations, we currently prepare a single arm phase II study to test the safety (GVHD) and efficacy of AA supplementation after allogeneic stem cell transplantation. This study is likely to provide important insights in how this vitamin functions in a complex, diseased organism and what groups of patients may benefit from this safe supplement.

Author Contributions: Gwendolyn N. Y. van Gorkom wrote the paper, Roel G. J. Klein Wolterink, Catharina H. M. J. Van Elssen, Lotte Wieten, Wilfred T. V. Germeraad and Gerard M. J. Bos were involved in the previous work on this topic in our group and/or substantively revised it.

Conflicts of Interest: The authors declare no conflict of interest.

References

1. Cameron, E.; Pauling, L. Supplemental ascorbate in the supportive treatment of cancer: Prolongation of survival times in terminal human cancer. *Proc. Natl. Acad. Sci. USA* **1976**, *73*, 3685–3689. [CrossRef] [PubMed]
2. Creagan, E.T.; Moertel, C.G.; O'Fallon, J.R.; Schutt, A.J.; O'Connell, M.J.; Rubin, J.; Frytak, S. Failure of high-dose vitamin C (ascorbic acid) therapy to benefit patients with advanced cancer. A controlled trial. *N. Engl. J. Med.* **1979**, *301*, 687–690. [CrossRef] [PubMed]
3. Moertel, C.G.; Fleming, T.R.; Creagan, E.T.; Rubin, J.; O'Connell, M.J.; Ames, M.M. High-dose vitamin C versus placebo in the treatment of patients with advanced cancer who have had no prior chemotherapy. A randomized double-blind comparison. *N. Engl. J. Med.* **1985**, *312*, 137–141. [CrossRef] [PubMed]
4. Young, J.I.; Zuchner, S.; Wang, G. Regulation of the Epigenome by Vitamin C. *Annu. Rev. Nutr.* **2015**, *35*, 545–564. [CrossRef] [PubMed]
5. Harrison, F.E.; Meredith, M.E.; Dawes, S.M.; Saskowski, J.L.; May, J.M. Low ascorbic acid and increased oxidative stress in gulo($-/-$) mice during development. *Brain Res.* **2010**, *1349*, 143–152. [CrossRef] [PubMed]
6. Uchio, R.; Hirose, Y.; Murosaki, S.; Yamamoto, Y.; Ishigami, A. High dietary intake of vitamin C suppresses age-related thymic atrophy and contributes to the maintenance of immune cells in vitamin C-deficient senescence marker protein-30 knockout mice. *Br. J. Nutr.* **2015**, *113*, 603–609. [CrossRef] [PubMed]
7. Carr, A.C.; Maggini, S. Vitamin C and Immune Function. *Nutrients* **2017**, *9*, 1211. [CrossRef] [PubMed]
8. Wilson, J.X. Regulation of vitamin C transport. *Annu. Rev. Nutr.* **2005**, *25*, 105–125. [CrossRef] [PubMed]
9. Omaye, S.T.; Schaus, E.E.; Kutnink, M.A.; Hawkes, W.C. Measurement of vitamin C in blood components by high-performance liquid chromatography. Implication in assessing vitamin C status. *Ann. N. Y. Acad. Sci.* **1987**, *498*, 389–401. [CrossRef] [PubMed]
10. Evans, R.M.; Currie, L.; Campbell, A. The distribution of ascorbic acid between various cellular components of blood, in normal individuals, and its relation to the plasma concentration. *Br. J. Nutr.* **1982**, *47*, 473–482. [CrossRef] [PubMed]
11. Roux, E.; Dumont-Girard, F.; Starobinski, M.; Siegrist, C.A.; Helg, C.; Chapuis, B.; Roosnek, E. Recovery of immune reactivity after T-cell-depleted bone marrow transplantation depends on thymic activity. *Blood* **2000**, *96*, 2299–2303. [PubMed]
12. Bosch, M.; Khan, F.M.; Storek, J. Immune reconstitution after hematopoietic cell transplantation. *Curr. Opin. Hematol.* **2012**, *19*, 324–335. [CrossRef] [PubMed]
13. Huijskens, M.J.; Walczak, M.; Koller, N.; Briede, J.J.; Senden-Gijsbers, B.L.; Schnijderberg, M.C.; Bos, G.M.; Germeraad, W.T. Technical advance: Ascorbic acid induces development of double-positive T cells from human hematopoietic stem cells in the absence of stromal cells. *J. Leukoc. Biol.* **2014**, *96*, 1165–1175. [CrossRef] [PubMed]
14. Huijskens, M.J.; Walczak, M.; Sarkar, S.; Atrafi, F.; Senden-Gijsbers, B.L.; Tilanus, M.G.; Bos, G.M.; Wieten, L.; Germeraad, W.T. Ascorbic acid promotes proliferation of natural killer cell populations in culture systems applicable for natural killer cell therapy. *Cytotherapy* **2015**, *17*, 613–620. [CrossRef] [PubMed]
15. Huijskens, M.J.; Wodzig, W.K.; Walczak, M.; Germeraad, W.T.; Bos, G.M. Ascorbic acid serum levels are reduced in patients with hematological malignancies. *Results Immunol.* **2016**, *6*, 8–10. [CrossRef] [PubMed]
16. Jenkinson, E.J.; Anderson, G.; Owen, J.J. Studies on T cell maturation on defined thymic stromal cell populations in vitro. *J. Exp. Med.* **1992**, *176*, 845–853. [CrossRef] [PubMed]
17. Schmitt, T.M.; Zuniga-Pflucker, J.C. Induction of T cell development from hematopoietic progenitor cells by delta-like-1 in vitro. *Immunity* **2002**, *17*, 749–756. [CrossRef]
18. Meek, B.; Cloosen, S.; Borsotti, C.; Van Elssen, C.H.; Vanderlocht, J.; Schnijderberg, M.C.; van der Poel, M.W.; Leewis, B.; Hesselink, R.; Manz, M.G.; et al. In vitro-differentiated T/natural killer-cell progenitors derived from human CD34+ cells mature in the thymus. *Blood* **2010**, *115*, 261–264. [CrossRef] [PubMed]
19. Manning, J.; Mitchell, B.; Appadurai, D.A.; Shakya, A.; Pierce, L.J.; Wang, H.; Nganga, V.; Swanson, P.C.; May, J.M.; Tantin, D.; et al. Vitamin C promotes maturation of T-cells. *Antioxid. Redox Signal.* **2013**, *19*, 2054–2067. [CrossRef] [PubMed]
20. Campbell, J.D.; Cole, M.; Bunditrutavorn, B.; Vella, A.T. Ascorbic acid is a potent inhibitor of various forms of T cell apoptosis. *Cell. Immunol.* **1999**, *194*, 1–5. [CrossRef] [PubMed]

21. Maeng, H.G.; Lim, H.; Jeong, Y.J.; Woo, A.; Kang, J.S.; Lee, W.J.; Hwang, Y.I. Vitamin C enters mouse T cells as dehydroascorbic acid in vitro and does not recapitulate in vivo vitamin C effects. *Immunobiology* **2009**, *214*, 311–320. [CrossRef] [PubMed]

22. Eylar, E.; Baez, I.; Navas, J.; Mercado, C. Sustained levels of ascorbic acid are toxic and immunosuppressive for human T cells. *P. R. Health Sci. J.* **1996**, *15*, 21–26. [PubMed]

23. Hong, J.M.; Kim, J.H.; Kang, J.S.; Lee, W.J.; Hwang, Y.I. Vitamin C is taken up by human T cells via sodium-dependent vitamin C transporter 2 (SVCT2) and exerts inhibitory effects on the activation of these cells in vitro. *Anat. Cell Biol.* **2016**, *49*, 88–98. [CrossRef] [PubMed]

24. Fraser, R.C.; Pavlovic, S.; Kurahara, C.G.; Murata, A.; Peterson, N.S.; Taylor, K.B.; Feigen, G.A. The effect of variations in vitamin C intake on the cellular immune response of guinea pigs. *Am. J. Clin. Nutr.* **1980**, *33*, 839–847. [CrossRef] [PubMed]

25. Badr, G.; Bashandy, S.; Ebaid, H.; Mohany, M.; Sayed, D. Vitamin C supplementation reconstitutes polyfunctional T cells in streptozotocin-induced diabetic rats. *Eur. J. Nutr.* **2012**, *51*, 623–633. [CrossRef] [PubMed]

26. Kennes, B.; Dumont, I.; Brohee, D.; Hubert, C.; Neve, P. Effect of vitamin C supplements on cell-mediated immunity in old people. *Gerontology* **1983**, *29*, 305–310. [CrossRef] [PubMed]

27. Kay, N.E.; Holloway, D.E.; Hutton, S.W.; Bone, N.D.; Duane, W.C. Human T-cell function in experimental ascorbic acid deficiency and spontaneous scurvy. *Am. J. Clin. Nutr.* **1982**, *36*, 127–130. [CrossRef] [PubMed]

28. Noh, K.; Lim, H.; Moon, S.K.; Kang, J.S.; Lee, W.J.; Lee, D.; Hwang, Y.I. Mega-dose Vitamin C modulates T cell functions in Balb/c mice only when administered during T cell activation. *Immunol. Lett.* **2005**, *98*, 63–72. [CrossRef] [PubMed]

29. Chang, H.-H.; Chen, C.; Lin, J.-Y. High dose vitamin C supplementation increases the Th1/Th2 cytokine secretion ratio, but decreases eosinophilic infiltration in bronchoalveolar lavage fluid of ovalbumin-sensitized and challenged mice. *J. Agric. Food Chem.* **2009**, *57*, 10471–10476. [CrossRef] [PubMed]

30. Jeong, Y.J.; Hong, S.W.; Kim, J.H.; Jin, D.H.; Kang, J.S.; Lee, W.J.; Hwang, Y.I. Vitamin C-treated murine bone marrow-derived dendritic cells preferentially drive naive T cells into Th1 cells by increased IL-12 secretions. *Cell. Immunol.* **2011**, *266*, 192–199. [CrossRef] [PubMed]

31. Song, M.H.; Nair, V.S.; Oh, K.I. Vitamin C enhances the expression of IL17 in a Jmjd2-dependent manner. *BMB Rep.* **2017**, *50*, 49–54. [CrossRef] [PubMed]

32. Jeong, Y.J.; Kim, J.H.; Hong, J.M.; Kang, J.S.; Kim, H.R.; Lee, W.J.; Hwang, Y.I. Vitamin C treatment of mouse bone marrow-derived dendritic cells enhanced CD8(+) memory T cell production capacity of these cells in vivo. *Immunobiology* **2014**, *219*, 554–564. [CrossRef] [PubMed]

33. Sasidharan Nair, V.; Song, M.H.; Oh, K.I. Vitamin C Facilitates Demethylation of the Foxp3 Enhancer in a Tet-Dependent Manner. *J. Immunol.* **2016**, *196*, 2119–2131. [CrossRef] [PubMed]

34. Yue, X.; Trifari, S.; Aijo, T.; Tsagaratou, A.; Pastor, W.A.; Zepeda-Martinez, J.A.; Lio, C.W.; Li, X.; Huang, Y.; Vijayanand, P.; et al. Control of Foxp3 stability through modulation of TET activity. *J. Exp. Med.* **2016**, *213*, 377–397. [CrossRef] [PubMed]

35. Tahiliani, M.; Koh, K.P.; Shen, Y.; Pastor, W.A.; Bandukwala, H.; Brudno, Y.; Agarwal, S.; Iyer, L.M.; Liu, D.R.; Aravind, L.; et al. Conversion of 5-methylcytosine to 5-hydroxymethylcytosine in mammalian DNA by MLL partner TET1. *Science* **2009**, *324*, 930–935. [CrossRef] [PubMed]

36. Ito, S.; D'Alessio, A.C.; Taranova, O.V.; Hong, K.; Sowers, L.C.; Zhang, Y. Role of Tet proteins in 5mC to 5hmC conversion, ES-cell self-renewal and inner cell mass specification. *Nature* **2010**, *466*, 1129–1133. [CrossRef] [PubMed]

37. Blaschke, K.; Ebata, K.T.; Karimi, M.M.; Zepeda-Martinez, J.A.; Goyal, P.; Mahapatra, S.; Tam, A.; Laird, D.J.; Hirst, M.; Rao, A.; et al. Vitamin C induces Tet-dependent DNA demethylation and a blastocyst-like state in ES cells. *Nature* **2013**, *500*, 222–226. [CrossRef] [PubMed]

38. Chung, T.L.; Brena, R.M.; Kolle, G.; Grimmond, S.M.; Berman, B.P.; Laird, P.W.; Pera, M.F.; Wolvetang, E.J. Vitamin C promotes widespread yet specific DNA demethylation of the epigenome in human embryonic stem cells. *Stem Cells* **2010**, *28*, 1848–1855. [CrossRef] [PubMed]

39. Nikolouli, E.; Hardtke-Wolenski, M.; Hapke, M.; Beckstette, M.; Geffers, R.; Floess, S.; Jaeckel, E.; Huehn, J. Alloantigen-Induced Regulatory T Cells Generated in Presence of Vitamin C Display Enhanced Stability of Foxp3 Expression and Promote Skin Allograft Acceptance. *Front. Immunol.* **2017**, *8*, 748. [CrossRef] [PubMed]

40. Kasahara, H.; Kondo, T.; Nakatsukasa, H.; Chikuma, S.; Ito, M.; Ando, M.; Kurebayashi, Y.; Sekiya, T.; Yamada, T.; Okamoto, S.; et al. Generation of allo-antigen-specific induced Treg stabilized by vitamin C treatment and its application for prevention of acute graft versus host disease model. *Int. Immunol.* **2017**, *29*, 457–469. [CrossRef] [PubMed]

41. Gao, Y.; Lu, B.; Zhai, J.; Liu, Y.; Qi, H.; Yao, Y.; Chai, Y.; Shou, S. The Parenteral Vitamin C Improves Sepsis and Sepsis-Induced Multiple Organ Dysfunction Syndrome via Preventing Cellular Immunosuppression. *Mediat. Inflamm.* **2017**, *2017*, 4024672. [CrossRef] [PubMed]

42. Ichiyama, K.; Mitsuzumi, H.; Zhong, M.; Tai, A.; Tsuchioka, A.; Kawai, S.; Yamamoto, I.; Gohda, E. Promotion of IL-4- and IL-5-dependent differentiation of anti-μ-primed B cells by ascorbic acid 2-glucoside. *Immunol. Lett.* **2009**, *122*, 219–226. [CrossRef] [PubMed]

43. Woo, A.; Kim, J.H.; Jeong, Y.J.; Maeng, H.G.; Lee, Y.T.; Kang, J.S.; Lee, W.J.; Hwang, Y.I. Vitamin C acts indirectly to modulate isotype switching in mouse B cells. *Anat. Cell Biol.* **2010**, *43*, 25–35. [CrossRef] [PubMed]

44. Tanaka, M.; Muto, N.; Gohda, E.; Yamamoto, I. Enhancement by ascorbic acid 2-glucoside or repeated additions of ascorbate of mitogen-induced IgM and IgG productions by human peripheral blood lymphocytes. *Jpn. J. Pharmacol.* **1994**, *66*, 451–456. [CrossRef] [PubMed]

45. Prinz, W.; Bloch, J.; Gilich, G.; Mitchell, G. A systematic study of the effect of vitamin C supplementation on the humoral immune response in ascorbate-dependent mammals. I. The antibody response to sheep red blood cells (a T-dependent antigen) in guinea pigs. *Int. J. Vitam. Nutr. Res.* **1980**, *50*, 294–300. [PubMed]

46. Feigen, G.A.; Smith, B.H.; Dix, C.E.; Flynn, C.J.; Peterson, N.S.; Rosenberg, L.T.; Pavlovic, S.; Leibovitz, B. Enhancement of antibody production and protection against systemic anaphylaxis by large doses of vitamin C. *Res. Commun. Chem. Pathol. Pharmacol.* **1982**, *38*, 313–333. [CrossRef]

47. Albers, R.; Bol, M.; Bleumink, R.; Willems, A.A.; Pieters, R.H. Effects of supplementation with vitamins A, C, and E, selenium, and zinc on immune function in a murine sensitization model. *Nutrition* **2003**, *19*, 940–946. [CrossRef]

48. Hesta, M.; Ottermans, C.; Krammer-Lukas, S.; Zentek, J.; Hellweg, P.; Buyse, J.; Janssens, G.P. The effect of vitamin C supplementation in healthy dogs on antioxidative capacity and immune parameters. *J. Anim. Physiol. Anim. Nutr.* **2009**, *93*, 26–34. [CrossRef] [PubMed]

49. Goodwin, J.S.; Garry, P.J. Relationship between megadose vitamin supplementation and immunological function in a healthy elderly population. *Clin. Exp. Immunol.* **1983**, *51*, 647–653. [PubMed]

50. Amakye-Anim, J.; Lin, T.L.; Hester, P.Y.; Thiagarajan, D.; Watkins, B.A.; Wu, C.C. Ascorbic acid supplementation improved antibody response to infectious bursal disease vaccination in chickens. *Poult. Sci.* **2000**, *79*, 680–688. [CrossRef] [PubMed]

51. Wu, C.C.; Dorairajan, T.; Lin, T.L. Effect of ascorbic acid supplementation on the immune response of chickens vaccinated and challenged with infectious bursal disease virus. *Vet. Immunol. Immunopathol.* **2000**, *74*, 145–152. [CrossRef]

52. Vallance, S. Relationships between ascorbic acid and serum proteins of the immune system. *Br. Med. J.* **1977**, *2*, 437–438. [CrossRef] [PubMed]

53. Prinz, W.; Bortz, R.; Bregin, B.; Hersch, M. The effect of ascorbic acid supplementation on some parameters of the human immunological defence system. *Int. J. Vitam. Nutr. Res.* **1977**, *47*, 248–257. [PubMed]

54. Anderson, R.; Oosthuizen, R.; Maritz, R.; Theron, A.; Van Rensburg, A.J. The effects of increasing weekly doses of ascorbate on certain cellular and humoral immune functions in normal volunteers. *Am. J. Clin. Nutr.* **1980**, *33*, 71–76. [CrossRef] [PubMed]

55. Huwyler, T.; Hirt, A.; Morell, A. Effect of ascorbic acid on human natural killer cells. *Immunol. Lett.* **1985**, *10*, 173–176. [CrossRef]

56. Toliopoulos, I.K.; Simos, Y.V.; Daskalou, T.A.; Verginadis, I.I.; Evangelou, A.M.; Karkabounas, S.C. Inhibition of platelet aggregation and immunomodulation of NK lymphocytes by administration of ascorbic acid. *Indian J. Exp. Biol.* **2011**, *49*, 904–908. [PubMed]

57. Vojdani, A.; Ghoneum, M. In vivo effect of ascorbic acid on enhancement of natural killer cell activity. *Nutr. Res.* **1993**, *13*, 753–764. [CrossRef]

58. Heuser, G.; Vojdani, A. Enhancement of natural killer cell activity and T and B cell function by buffered vitamin C in patients exposed to toxic chemicals: The role of protein kinase-C. *Immunopharmacol. Immunotoxicol.* **1997**, *19*, 291–312. [CrossRef] [PubMed]

59. Farmakis, D.; Giakoumis, A.; Polymeropoulos, E.; Aessopos, A. Pathogenetic aspects of immune deficiency associated with beta-thalassemia. *Med. Sci. Monit.* **2003**, *9*, Ra19–R22. [PubMed]
60. Atasever, B.; Ertan, N.Z.; Erdem-Kuruca, S.; Karakas, Z. In vitro effects of vitamin C and selenium on NK activity of patients with beta-thalassemia major. *Pediatr. Hematol. Oncol.* **2006**, *23*, 187–197. [CrossRef] [PubMed]
61. Lynch, S.R.; Cook, J.D. Interaction of vitamin C and iron. *Ann. N. Y. Acad. Sci.* **1980**, *355*, 32–44. [CrossRef] [PubMed]
62. Hua, Y.; Wang, C.; Jiang, H.; Wang, Y.; Liu, C.; Li, L.; Liu, H.; Shao, Z.; Fu, R. Iron overload may promote alteration of NK cells and hematopoietic stem/progenitor cells by JNK and P38 pathway in myelodysplastic syndromes. *Int. J. Hematol.* **2017**, *106*, 248–257. [CrossRef] [PubMed]
63. Kim, J.E.; Cho, H.S.; Yang, H.S.; Jung, D.J.; Hong, S.W.; Hung, C.F.; Lee, W.J.; Kim, D. Depletion of ascorbic acid impairs NK cell activity against ovarian cancer in a mouse model. *Immunobiology* **2012**, *217*, 873–881. [CrossRef] [PubMed]
64. Munneke, J.M.; Bjorklund, A.T.; Mjosberg, J.M.; Garming-Legert, K.; Bernink, J.H.; Blom, B.; Huisman, C.; van Oers, M.H.; Spits, H.; Malmberg, K.J.; et al. Activated innate lymphoid cells are associated with a reduced susceptibility to graft-versus-host disease. *Blood* **2014**, *124*, 812–821. [CrossRef] [PubMed]
65. Vacca, P.; Montaldo, E.; Croxatto, D.; Moretta, F.; Bertaina, A.; Vitale, C.; Locatelli, F.; Mingari, M.C.; Moretta, L. NK Cells and Other Innate Lymphoid Cells in Hematopoietic Stem Cell Transplantation. *Front. Immunol.* **2016**, *7*, 188. [CrossRef] [PubMed]

antioxidants

MDPI

Article

Attenuation of Red Blood Cell Storage Lesions with Vitamin C

Kimberly Sanford [1], Bernard J. Fisher [2], Evan Fowler [2], Alpha A. Fowler III [2] and
Ramesh Natarajan [2,*]

[1]   Department of Pathology, Director, Transfusion Services, Virginia Commonwealth University,
      Richmond, VA 23298, USA; kimberly.sanford@vcuhealth.org
[2]   Department of Internal Medicine, Virginia Commonwealth University, Richmond, VA 23298, USA;
      bernard.fisher@vcuhealth.org (B.J.F.); evan.fowler@vcuhealth.org (E.F.); alpha.fowler@vcuhealth.org (A.A.F.)
*   Correspondence: ramesh.natarajan@vcuhealth.org; Tel.: +1-804-827-1013

Received: 6 June 2017; Accepted: 8 July 2017; Published: 12 July 2017

Abstract: Stored red blood cells (RBCs) undergo oxidative stress that induces deleterious metabolic, structural, biochemical, and molecular changes collectively referred to as "storage lesions". We hypothesized that vitamin C (VitC, reduced or oxidized) would reduce red cell storage lesions, thus prolonging their storage duration. Whole-blood-derived, leuko-reduced, SAGM (saline-adenine-glucose-mannitol)-preserved RBC concentrates were equally divided into four pediatric storage bags and the following additions made: (1) saline (saline); (2) 0.3 mmol/L reduced VitC (Lo VitC); (3) 3 mmol/L reduced VitC (Hi VitC); or (4) 0.3 mmol/L oxidized VitC (dehydroascorbic acid, DHA) as final concentrations. Biochemical and rheological parameters were serially assessed at baseline (prior to supplementation) and Days 7, 21, 42, and 56 for RBC VitC concentration, pH, osmotic fragility by mechanical fragility index, and percent hemolysis, LDH release, glutathione depletion, RBC membrane integrity by scanning electron microscopy, and Western blot for β-spectrin. VitC exposure (reduced and oxidized) significantly increased RBC antioxidant status with varying dynamics and produced trends in reduction in osmotic fragility and increases in membrane integrity. Conclusion: VitC partially protects RBC from oxidative changes during storage. Combining VitC with other antioxidants has the potential to improve long-term storage of RBC.

Keywords: vitamin C; RBC storage lesions; mean fragility index; β-spectrin; scanning electron microscopy

1. Introduction

Red blood cell (RBC) transfusion is a life-saving procedure whose primary objective is to sustain tissue and organ oxygenation in case of acute anemia due to massive hemorrhage or chronic anemia secondary to bone marrow dysfunction. RBC units, prepared by plasma removal and leukocyte depletion, are stored for 42 days [1]. This extension of RBC shelf-life is possible due to the combined application of preparation methods, suitable storage additive solutions (CPD—citrate phosphate-dextrose, saline-adenine-glucose-mannitol (SAGM) or Optisol (AS-5)), polyvinyl chloride blood bags, and storage at 4 °C [2]. However, despite these measures, stored RBCs undergo deleterious metabolic, structural, biochemical, and molecular changes collectively referred to as "storage lesions" [3]. Storage lesions are characterized by ATP depletion, loss of 2,3-diphosphoglycerate (2,3-DPG), glutathione (GSH), and nicotinamide adenine dinucleotide (NADH/NADPH) depletion with subsequent oxidation of hemoglobin, exhaustion of the endogenous antioxidants, leakage of lactate, lactate dehydrogenase (LDH), hemoglobin, and potassium ions into the suspending medium. Membrane micro-vesiculation as well as reversible and irreversible morphological changes also occur [4–8].

A major contributor to storage lesions is oxidative stress that progressively increases over the storage period due to consumption of endogenous anti-oxidants. Wither and colleagues identified oxidative modifications of functional residues of the hemoglobin beta chain [9]. Rinalducci and colleagues detected progressive linkage of cytosolic proteins to the cell membrane that included antioxidant and metabolic enzymes. Their detailed analysis of aged RBC membranes suggested that peroxiredoxin-2 could serve as a marker of oxidative stress [10]. Studies using mass spectrometry and electron microscopy have shown that various factors including altered cation homeostasis, reprogrammed energy, and redox metabolism contribute to the progressive accumulation of oxidative stress and storage lesions in stored RBCs [8]. Oxidative stress, in turn, promotes oxidative lesions to proteins (carbonylation, fragmentation, denatured hemoglobin) and lipids (peroxidation), which then contribute to the altered physiology of stored RBCs [8,11,12].

Pharmacological concentrations of VitC induces oxidative stress and GSH depletion while promoting increased glucose flux through the oxidative pentose phosphate pathway (PPP) in erythrocytes. Zhang and colleagues showed that erythrocytes incubated with VitC exhibit hemolysis [13]. Hemolysis was intensified in erythrocytes obtained from glucose-6-phosphate dehydrogenase (G6PD) deficient patients. Of importance in this work, erythrocytes were salvaged in Zhang's work and in reports by others [14] by the addition of antioxidants, by alterations of pH [15] or by addition of alkaline CPD, which preserves erythrocyte 2,3-DPG. [16].

RBC deformability is critical to RBC function, especially in the microvasculature where impaired deformability adversely affects capillary perfusion [17]. Oxidative injury also induces formation of RBC membrane microparticles with release of bioactive lipids that promote pro-inflammatory/pro-coagulant responses in the recipient, resulting in transfusion-related acute lung injury (TRALI) [18]. Another component of storage lesions is increased sub-lethal injury. Sub-lethal injury refers to the damage inflicted upon cells that does not result in their immediate hemolysis and that is exacerbated by oxidative stress [19,20]. It is expressed using the mechanical fragility index (MFI). Higher MFI values correspond to greater amounts of sub-lethal injury.

The addition of antioxidants to stored RBCs has the potential to mitigate oxidative injury and thus storage lesions. Recent studies indicate a renewed interest in the use of pharmacologic doses of vitamin C (VitC) to stored RBCs. Raval et al. showed that VitC exposure significantly reduced mechanical fragility and hemolysis in stored RBCs [19]. Stowell et al. demonstrated that the addition of an ascorbic acid solution to stored murine RBCs improved post-transfusion recovery while decreasing microparticle formation and allo-immunization [21]. Czubak et al. used a combination of sodium ascorbate and trolox to inhibit hemolysis and lipid peroxidation, while enhancing total antioxidant status [22]. These studies indicate that RBC preservation using optimal concentrations of VitC may prevent storage lesions.

We investigated the effects of reduced VitC (ascorbic acid: 0.3 mmol/L, 3 mmol/L) and oxidized VitC (dehydroascorbic acid [DHA], 0.3 mmol/L) on oxidative stress mediated changes in RBC's stored at 4 °C for 56 days. We hypothesized that VitC would reduce red cell storage lesions by attenuating osmotic fragility, LDH release, and glutathione depletion by maintaining RBC membrane integrity and structure, thus prolonging their storage duration, and improving their viability and function upon transfusion.

2. Materials and Methods

2.1. Research Involving Human Subjects

No human subjects were used for this study. However, freshly donated whole blood units from 5 de-identified community volunteer donors that meet the specific donor criteria established by the Food and Drug Administration through the Code of Federal Regulations (CFR), guidance documents and the American Association of Blood Banks (AABB) established professional standards for donor selection were purchased from Virginia Blood Services (Richmond, VA, USA). We declare that this study was carried out in accordance with the Declaration of Helsinki. Further, since no human subject information was gathered, approval by an institutional review board was not required.

2.2. Preparation and Storage of RBCs

Freshly donated whole blood units from 5 de-identified donors was purchased from Virginia Blood Services (Richmond, VA, USA). Blood was screened and RBC concentrates were prepared by Virginia Blood Services using standardized protocols of plasma removal and leukocyte depletion, followed by the addition of SAGM to RBCs. Each RBC unit was equally divided into four pediatric storage bags (volume of ~75 mL per aliquot) and supplementations for the study were made to each aliquot at the supplier facility as described below using aseptic technique. Prior to supplementation, an initial baseline sample was collected at the blood supplier facility and transported to the laboratory for analysis. RBCs that passed standard screening tests were transported two days later to the Virginia Commonwealth University Transfusion Medicine Center and stored at 4 °C for 56 days. All subsequent collections were made at the Virginia Commonwealth University Transfusion Medicine Center on Days 7, 21, 42 and 56.

2.3. Experimental Design and Study Groups

RBCs in pediatric storage bags were treated with one of four additives: (1) normal saline (saline); (2) 0.3 mmol/L reduced VitC (Lo VitC); (3) 3 mmol/L reduced VitC (Hi VitC); or (4) 0.3 mmol/L oxidized VitC (DHA) as final concentrations and gently mixed. The reduced VitC additive was preservative-free buffered ascorbic acid in water (Ascor L500, McGuff Pharmaceuticals, Santa Ana, CA, USA), pH 5.5–7.0 adjusted with sodium bicarbonate and sodium hydroxide. Oxidized VitC (DHA) was procured from Sigma-Aldrich (St. Louis, MO, USA). DHA was dissolved in saline at 60 °C for 30 min. It was filter sterilized and injected into the pediatric bags using a sterile technique.

2.4. Determination of RBC Vitamin C Content and pH

For vitamin C quantification, RBCs were pelleted by centrifugation (10,000 g, 10 min at 4 °C) and the supernatant plasma saved for pH determination. Pelleted RBCs were washed with cold saline; deproteinized in 200 µL of cold 20% trichloroacetic acid (TCA) followed by the addition of 200 µL of cold 0.2% dithiothreitol (DTT) to prevent oxidation. RBC lysates were vortexed and centrifuged at 10,000 g for 10 min at 4 °C. The supernatants were stored at −80 °C for batch vitamin C analysis. Total vitamin C content was assessed using a Tempol-OPDA-based fluorescence end-point assay as previously described [23].

2.5. Mechanical Fragility Test and Percent Hemolysis

The mechanical fragility test and percent hemolysis were performed on each sample as described by Raval et al. [19]. Briefly, at each time point, aliquots from each of the treatments were removed, adjusted to a standard hematocrit (Hct) of 40% with Dulbecco's phosphate buffered saline (DPBS with calcium and magnesium, GE Healthcare Life Sciences, HyClone Laboratories, Logan, UT, USA) and the hemoglobin (Hb) concentration determined spectrophotometrically (ABX Micros 60; Horiba, Ltd., Kyoto, Japan). Two milliliters of each aliquot were then added to each of five tubes (6 mL, 13 × 100 mm Vacuette blood collection tubes, Greiner Bio-One; Monroe, NC, USA), three of which contained five 3.2 mm steel ball bearings (BNMX-2, Type 316 balls; Small Parts, Inc., Miami Lakes, FL, USA) and two of which did not. The tubes with ball bearings were rocked on a rocker platform for 1 h, while the remaining tubes without bearings were not rocked and served as controls to ascertain the initial concentration of free Hb in each aliquot. After rocking, all tubes were centrifuged twice, and the free Hb concentrations in the supernatants determined spectrophotometrically by light absorbance at 540 nm (Bio-Rad SmartSpecTM 3000, Bio-Rad Inc., Hercules, CA, USA). The MFI and percent hemolysis were calculated as previously described by Raval et al. [19] using the following equations.

MFI calculation:

$$\frac{(fHb_{rocked} - fHb_{control})}{(Hb_{aliquot} - fHb_{control})} \times 100$$

Percent Hemolysis Equation:

$$\frac{(100 - Hct\ of\ sample) \times fHb_{control}}{Hb_{aliquot}}$$

$fHb_{control}$ = mean free Hb concentration in the supernatant of the unrocked control sample;
fHb_{rocked} = mean free Hb concentration in the supernatant of the rocked sample;
$Hb_{aliquot}$ = mean total Hb concentration of the RBC aliquot at an Hct of 40%.

2.6. Measurement of LDH Activity

LDH was measured using a Lactate Dehydrogenase Activity Assay Kit (Sigma-Aldrich, St. Louis, MO, USA) according to manufacturer's instructions. An aliquot of RBC concentrate was frozen at $-80\ °C$ for batch analysis. In addition, an aliquot of the RBC concentrate was pelleted by centrifugation ($10,000\ g$, 10 min at $4\ °C$) and the supernatant (plasma) stored at $-80\ °C$. Thawed RBC samples were rapidly homogenized on ice in LDH Assay Buffer and insoluble material removed by centrifugation ($10,000\ g$, 15 min at $4\ °C$). The soluble fraction was used for assay while plasma from RBC concentrate was assayed directly.

2.7. Measurement of Glutathione Content of RBC

Glutathione was measured using a Glutathione Assay Kit (Cayman Chemical, Ann Arbor, MI, USA) according to manufacturer's instructions. An aliquot of RBC was pelleted by centrifugation ($10,000\ g$, 10 min at $4\ °C$) and lysed in 4 volumes of cold water. After centrifugation ($10,000\ g$, 15 min at $4\ °C$) the erythrocyte lysate was deproteinized with an equal volume of 10% metaphosphoric acid and centrifuged ($10,000\ g$, 10 min at $4\ °C$). The supernatants were stored at $-80\ °C$ for batch analysis.

2.8. Western Blot for β-Spectrin

RBCs were pelleted by centrifugation ($10,000\ g$, 10 min at $4\ °C$) and washed with cold PBS. RBCs were lysed with RIPA buffer containing protease and phosphatase inhibitors (Sigma-Aldrich, St. Louis, MO, USA). RBC lysates were resolved by SDS polyacrylamide gel electrophoresis (4–20%) and electrophoretically transferred to polyvinylidene fluoride membranes (0.2 μm pore size) according to manufacturer's instructions (Life Technologies, Carlsbad, CA, USA). Immunodetection was performed using chemiluminescence with the Renaissance Western Blot Chemiluminescence Reagent Plus (Perkin Elmer Life Sciences Inc., Boston, MA, USA). Blots were stripped using the Restore™ Western Blot Stripping Buffer (Pierce Biotechnology Inc., Rockford, IL, USA) as described by the manufacturer. Purified rabbit monoclonal antibodies to β-spectrin 1 (ab 129065, Abcam, Cambridge, MA, USA) and actin (sc-1616, Santa Cruz Biotechnology) were used in this study. Optical densities of antibody-specific bands were determined using Quantity One acquisition and analysis software (Bio-Rad, Hercules, CA, USA).

2.9. Scanning Electron Microscopy (SEM) of RBC

For SEM studies, RBCs were initially fixed in 0.2% glutaraldehyde in 0.1 M cacodylate buffer at room temperature for 15 min. Following centrifugation and removal of supernatant, RBCs were fixed in 1.5% glutaraldehyde in 0.1 M cacodylate buffer and stored at $4\ °C$ before further processing at the Virginia Commonwealth University Department of Neurobiology & Anatomy Microscopy Facility. Here, the fixative was replaced with distilled water, and mounted on poly-l-lysine-coated glass slides. The glass slides were kept in a moist atmosphere and dehydrated in graded ethanol (50%, 70%, 80%,

95% and 100%). Samples were allowed to air dry, mounted on stubs with silver paint, dried overnight, followed by several coats of gold sputter, and then visualized using a Zeiss Evo 50XVP microscope (Carl Zeiss SMT, Inc., Peabody, MA, USA). RBC shapes were identified using the classification postulated by Bessis [24]. The percentages of discocytes, echinocytes, spheroechinocytes, stomatocytes, spherostomatocytes, and spherocytes were evaluated by counting 500 cells in randomly chosen fields. Reversible and irreversible shapes were determined as previously described [25]. RBC manifesting echinocyte and stomatocyte shapes are traditionally considered to be able to return to the discocyte shape under certain conditions. Thus, these RBC shape changes are considered potentially reversible transformations. In contrast, RBC assuming spheroechinocyte, spherostomatocyte, spherocyte, ovalocyte, and degenerated shapes are considered irreversibly changed cells.

2.10. Statistical Analysis

Statistical analysis was performed using SAS 9.4 and GraphPad Prism 7.0 (GraphPad Software, San Diego, CA, USA). Data are expressed as mean ± SE. Results were compared by one-way ANOVA and the post-hoc Tukey test to identify specific differences between groups. Statistical significance was confirmed at a *p*-value of < 0.05.

3. Results

3.1. Intracellular VitC Concentrations in RBC

Measurement of intracellular VitC content showed that RBC stored for 56 days at 4 °C in saline (absence of VitC) are gradually depleted of VitC. The addition of exogenous VitC resulted in a statistically significant increase in RBC VitC content (Figure 1). This was particularly evident with addition of Hi VitC (3 mmol/L, $p < 0.05$) and oxidized VitC (DHA, $p < 0.05$). Intracellular RBC VitC content continued to gradually increase with Lo ($p < 0.05$, on day 42 and 56) and Hi VitC ($p < 0.05$, all time-points), but did not change further after 7 days with DHA.

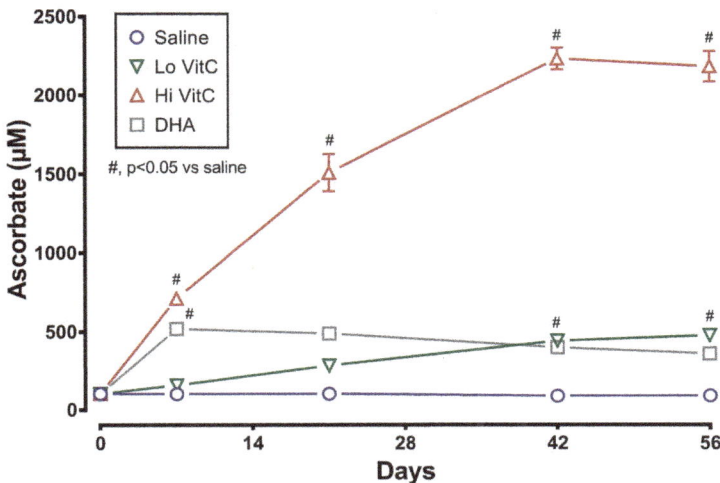

Figure 1. VitC exposure increased intracellular RBC ascorbate concentrations during storage. RBCs supplemented with reduced VitC (Lo/Hi) or oxidized VitC (DHA) had significantly higher intracellular levels of VitC compared to saline controls. Data are significant for Lo VitC on Days 42 and 56, for Hi VitC on Days 7, 21, 42, and 56, and for DHA on Day 7. ($N = 5$/group, $p < 0.05$).

3.2. Changes in pH

We determined the plasma pH of packed RBC stored for 56 days in saline or VitC. As seen in Figure 2, there was a sharp decline in the pH, which started at Day 7 and continued to drop over 56 days. The addition of VitC (reduced or oxidized) had no significant impact on the decrease in pH.

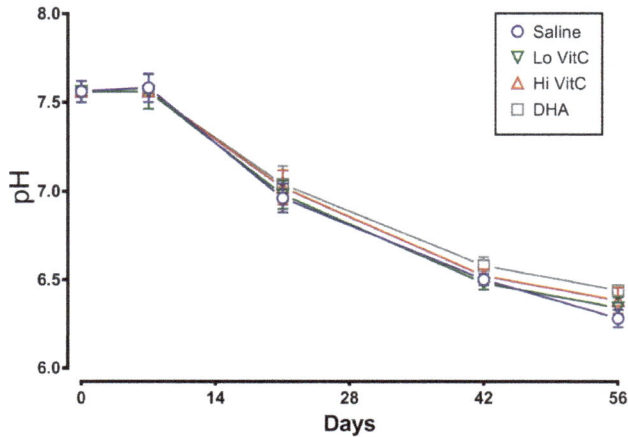

Figure 2. VitC exposure was associated with no changes in plasma pH. RBC concentrates showed a decline in plasma pH over 56 days of storage. The addition of reduced VitC (Lo/Hi) or oxidized VitC (DHA) did not alter the pH over the storage duration. ($N = 5$/group, $p > 0.05$).

3.3. Mean Fragility Index (MFI) and Percent Hemolysis

To determine whether addition of VitC preserves RBC membrane integrity, we examined MFI and percent hemolysis in RBC stored ± VitC for 56 days. As seen in Figure 3, the addition of VitC reduced the MFI of treated RBC. However, the decrease did not reach statistical significance ($p > 0.05$). VitC addition also did not significantly alter the percent hemolysis when compared to saline alone (Figure 4, $p > 0.05$).

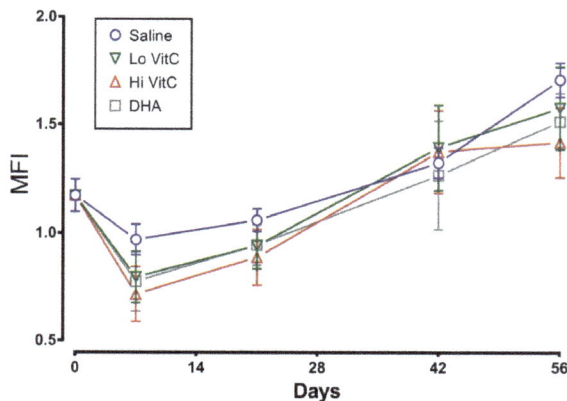

Figure 3. Effect of VitC exposure on the MFI of RBC. MFI of RBC treated with saline increased over 56 days of storage. The addition of reduced VitC (Lo/Hi) or oxidized VitC (DHA) reduced the MFI of treated RBC. However, this decrease did not reach statistical significance. ($N = 5$/group, $p > 0.05$).

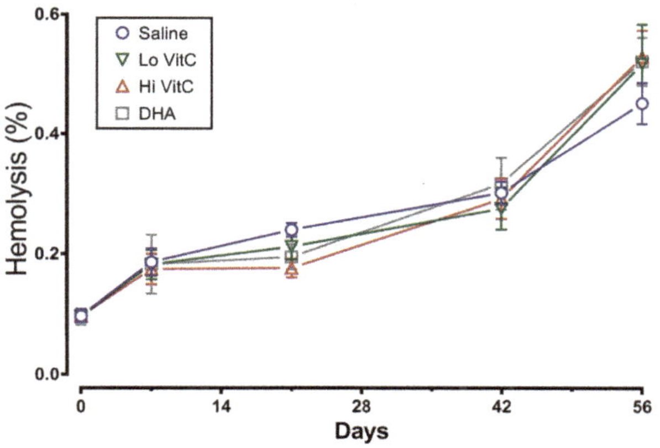

Figure 4. Effect of VitC exposure on percent hemolysis of RBC. The percent hemolysis of RBC treated with saline increased over 56 days of storage. The addition of reduced VitC (Lo/Hi) or oxidized VitC (DHA) did not significantly alter the percent hemolysis when compared to saline alone. ($N = 5$/group, $p > 0.05$).

3.4. Lactate Dehydrogenase Activity of Stored RBC

Another measure of membrane fragility is the release of LDH from damaged RBC. Therefore we measured LDH activity in stored RBC ± VitC for 56 days. As seen in Figure 5, LDH activity increased in the first 14 days and then plateaued out for the remaining 56 days. The addition of VitC did not significantly alter the release of LDH from stored RBC ($p > 0.05$).

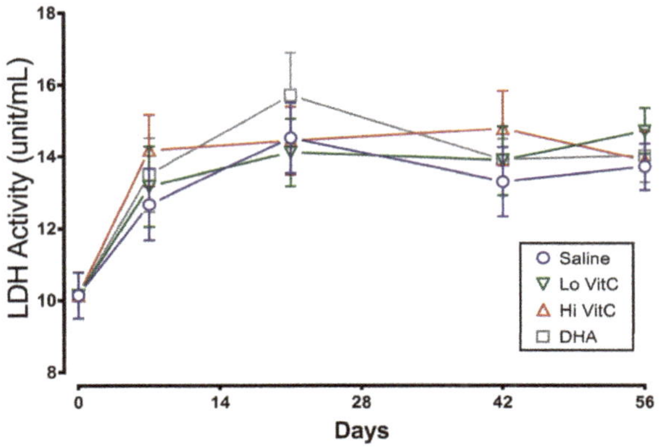

Figure 5. VitC exposure was associated with no changes in LDH activity. LDH activity in RBC treated with saline increased over the first 14 days and did not change further over the remaining 56 days. The addition of reduced VitC (Lo/Hi) or oxidized VitC (DHA) did not significantly alter the release of LDH from stored RBC. ($N = 5$/group, $p > 0.05$).

3.5. Glutathione Content of Stored RBC

We examined total GSH levels and the GSH/GSSG ratio in stored RBC ± VitC for 56 days. As seen in Figure 6a, total GSH content in RBC stored in saline decreased over 56 days, indicative of ongoing oxidative stress. However, the addition of VitC did not restore the GSH content in RBC. We also observed a slight decrease in the GSH/GSSG ratios on Day 7 with Hi-VitC and DHA-treated RBC (Figure 6b). However, this decrease was not statistically significant ($p > 0.05$).

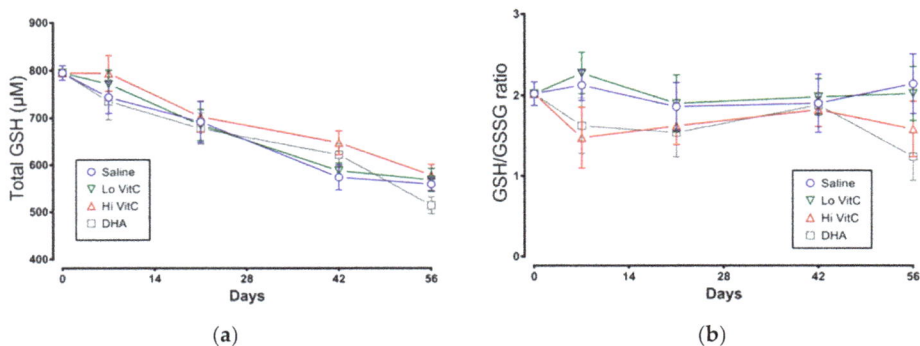

(**a**)   (**b**)

Figure 6. Effect of VitC exposure on glutathione content of stored RBC. (**a**) Total GSH content in RBC stored in saline decreased over 56 days. Addition of reduced VitC (Lo/Hi) or oxidized VitC (DHA) did not significantly alter the total GSH content from stored RBC. (**b**) There was a slight decrease in the GSH/GSSG ratios on Day 7 with Hi-VitC and DHA-treated RBC. However, this decrease was not statistically significant. ($N = 5$/group, $p > 0.05$).

3.6. Expression of β-Spectrin in Stored RBC

To determine β-spectrin levels in our study, we performed Western blot analysis of RBC ± VitC. As seen in Figure 7, β-spectrin content of RBC stored in saline progressively declined. The addition of VitC (oxidized or reduced) had a protective effect on β-spectrin levels. However, these changes did not reach statistical significance ($p > 0.05$).

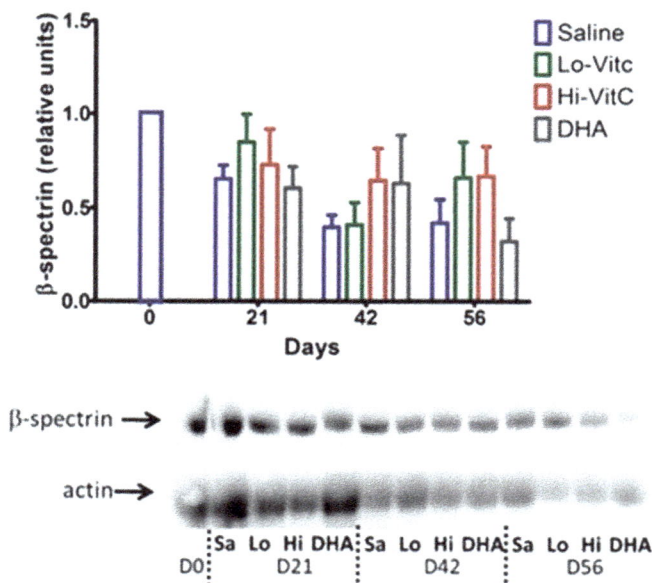

Figure 7. Expression of β-spectrin in stored RBC. The β-spectrin content of RBC stored in saline progressively declined over 56 days. Addition of reduced VitC (Lo/Hi) or oxidized VitC (DHA) increased β-spectrin levels of RBC over saline (Sa). However, these changes did not reach statistical significance. ($N = 5$/group, $p > 0.05$).

3.7. Electron Microscopy of Stored RBC

To further examine RBC structure and integrity, we performed scanning electron microscopy (SEM) of RBC ± VitC. Stored RBCs undergo considerable shape change from a normal smooth discoid shape (discocyte) to a spheroid shape with speculae (echinocyte). While some of these changes are reversible, others are irreversible, thus making the RBC dysfunctional [15,16]. In our studies, we found that the percent of normal RBCs were higher on Day 56 in RBC stored in VitC (Figure 8). However, these changes again did not reach statistical significance ($p > 0.05$).

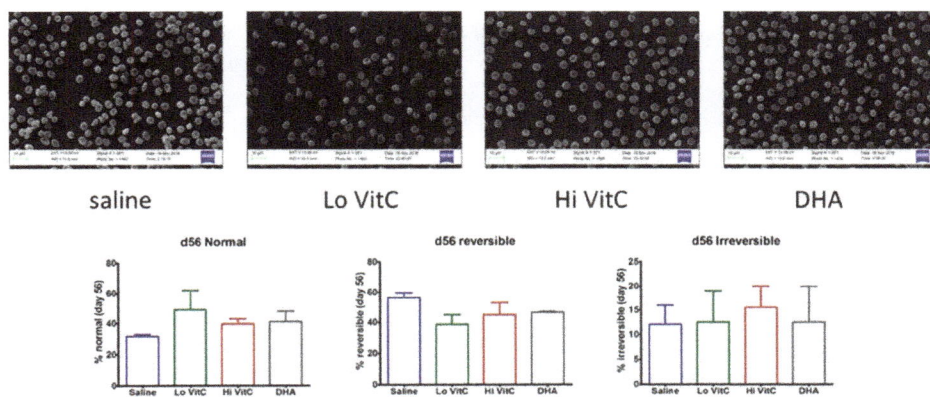

saline   Lo VitC   Hi VitC   DHA

Figure 8. Scanning electron microscopy of RBC. RBC stored in saline undergo considerable shape change from a normal smooth discoid shape (discocyte) to a spheroid shape with speculae (echinocyte) over 56 days with only 31% exhibiting a normal discocyte appearance on Day 56. The addition of reduced VitC (Lo/Hi) or oxidized VitC (DHA) increased the percent of normal discocyte shaped RBCs on Day 56. However, these changes again did not reach statistical significance. ($N = 3$/group, $p > 0.05$).

4. Discussion

Limited storage duration often results in a critical shortage of RBCs for routine/emergency use in emergency departments and critical care centers worldwide. Moreover, transfusion of these stored components to critically ill patients is often associated with increased multi-organ failure, nosocomial infections, ICU length of stay, and mortality [26,27]. Therefore, in this study, we examined the usefulness of the addition of VitC to stored RBCs. We hypothesized that VitC would reduce red blood cell storage lesions, thus prolonging their storage duration and improving their viability and function upon transfusion. While our results were promising, in that VitC addition showed beneficial trends towards better function without compromising the RBC's internal biochemical properties, many of the aspects analyzed did not reach statistical significance due to the relatively small sample size ($n = 5$) available for the study.

Most studies that have examined the use of VitC to improve RBC storage have used the reduced form of VitC, ascorbic acid. A unique aspect of our study was the use of both reduced VitC, ascorbic acid and oxidized VitC, dehydroascorbic acid. The rationale for using DHA was that RBC have receptors only for oxidized VitC and not for reduced VitC [28–30]. In agreement with observation, the addition of DHA resulted in a rapid uptake of VitC by RBC (Figure 1). In contrast, the addition of the Lo VitC dose (0.3 mmol/L) only produced gradual increases in intracellular RBC ascorbic acid content. However, the addition of Hi VitC (3 mmol/L) also produced an abrupt increase in RBC ascorbic acid content, which was sustained over 56 days (Figure 1). This is plausible only if ascorbic acid is oxidized to DHA, and the DHA is subsequently taken up via the glucose transporters. On this basis we conclude that there is ongoing oxidative stress in RBC stored for 56 days at 4 °C. Moreover, addition of reduced or oxidized VitC enhances RBC anti-oxidant levels by increasing their ascorbic acid content. However, in light of recent studies by Zhang et al., in which they showed that human erythrocytes express high levels of Glut1, take up DHA, and reduce it to VitC at the expense of reducing endogenous GSH, care must be employed with the use of DHA to increase RBC anti-oxidant stores [13].

Stored RBC are subject to oxidative stress induced storage lesions. Storage lesions deplete available adenine nucleotide pool (ATP, ADP, AMP) and 2,3-bisphosphoglyceric acid (2,3-DPG) levels resulting in limited oxygen unloading from hemoglobin upon systemic perfusion. A major factor causing the decrease of 2,3-DPG during storage is the fall in the pH of RBC. However, in agreement with Dawson et al. [31] and Raval et al. [19], addition of VitC did not negatively impact the pH of RBC (Figure 2). Moreover, the biochemical parameters of RBC units stored in VitC were similar to the saline control RBCs (Figures 3–6), demonstrating that VitC exposure did not adversely alter the biochemistry of RBCs at the dose utilized in these experiments.

Oxidative damage to RBC during storage alters cytoskeletal proteins and membrane phospholipids, resulting in increased osmotic fragility during storage and release of RBC contents into the supernatant plasma [11,12]. Others have previously shown that the MFI, a measure of RBC membrane sub-lethal injury, was found to be significantly lower in RBCs treated with 5.86 mmol/L or greater concentration of ascorbic acid by day 42 of storage [19]. Similarly, the percent hemolysis of ascorbic acid-treated RBCs was also significantly lower compared to the saline controls by Storage Day 21 when 8.78 mmol/L ascorbic acid was used [22]. In our study, using 0.3 mmol/L and 3 mmol/L VitC, we found similar beneficial trends in MFI and percent hemolysis (Figures 3 and 4). However, these changes did not reach statistical significance due to the smaller sample size of our study. The reason for using lower doses of VitC in our study when compared to others was to overcome the putative pro-oxidant activity of high concentrations of VitC that have been reported by some groups [32,33] to cause oxidative injury [31,34].

Total glutathione (GSH) content and the ratio of reduced over oxidized glutathione (GSH/GSSG) is an indicator of oxidative stress in stored RBC. In this study, addition of VitC did not alter the total GSH content or the GSH/GSSG ratio (Figure 6). Some studies have shown that, under physiological conditions, once DHA is taken up by cells, it is reduced intracellularly to VitC at the expense of GSH [35] or homocysteine [36]. However, the reverse reaction of VitC reducing oxidized GSH and restoring GSH levels have not been reported. Our data support those observations and suggest that while VitC can serve independently as a reducing agent against many reactive oxygen species (ROS) including superoxide anion (O_2^{-1}), hydroxyl radical ($OH\bullet^-$), singlet oxygen (O_2^*), and hypochlorous acid (HClO), it is incapable of donating electrons to oxidized GSH.

Tu et al. [37] previously showed that hyperglycemia in diabetes produced lower RBC ascorbate levels with increased RBC rigidity. Moreover, RBC β-spectrin expression correlated with its VitC concentrations [37] and that low β-spectrin levels were associated with reversible oxidative modification that affected RBC membrane integrity and structure. In agreement with Tu et al [37], we found that RBC storage over 56 days reduced β-spectrin levels (Figure 7). The addition of VitC (oxidized or reduced) potentially reversed these oxidative modifications and restored β-spectrin levels. This suggests that loading RBCs with VitC could restore membrane integrity and structure during storage. In agreement with these observations, using SEM we found that the percent of normal RBCs were higher on Day 56 in RBC stored in VitC compared to saline (Figure 8).

This study had several positives as well as limitations. The use of DHA, as well as the addition of VitC at the blood supplier facility, eliminated issues with transport that could have impacted the RBCs. A vast number of relevant assays were performed to determine change in RBC function over time. However, the biggest limitation was the sample size of the study, which severely impacted the statistical significance of the data. It is likely that increasing the sample size similar to that utilized by other groups [19,22] will yield statistically significant data. A second limitation was that generation of ROS was not measured directly, but was shown by indirect measures. Finally, it is possible that VitC alone may be insufficient to overcome the various aspects of RBC storage lesions. In line with this reasoning, Czubak et al. recently published a study where they showed that similar concentrations of VitC in combination with Trolox, a water soluble vitamin E analog, was required to reverse many RBC storage lesions [22].

5. Conclusions

In conclusion, these studies suggest that RBC stored for 56 days undergo oxidative modifications that alter their structure and function. VitC partially protects RBCs from these oxidative changes. Varying the dose of VitC, combining VitC with other antioxidants, and increasing the sample size used for this study is likely to provide more statistically significant observations.

Acknowledgments: Scanning electron microscopy was performed at the Virginia Commonwealth University Department of Neurobiology & Anatomy Microscopy Facility, supported, in part, with funding from NIH-NCI Cancer Center Support Grant P30 CA16059 and NIH-NCRR grant 1 S10 RR022495. This research was supported by grants from the Commonwealth Blood Foundation to Kimberly Sanford and Ramesh Natarajan.

Author Contributions: Ramesh Natarajan and Kimberly Sanford conceived and designed the experiments; Bernard J. Fisher, Evan Fowler and Ramesh Natarajan performed the experiments; Ramesh Natarajan and Bernard J. Fisher analyzed the data; Alpha A. Fowler III contributed reagents/materials/analysis tools; Alpha A. Fowler III, Bernard J. Fisher and Ramesh Natarajan wrote the paper.

Conflicts of Interest: The authors declare no conflict of interest. The founding sponsors had no role in the design of the study; in the collection, analyses, or interpretation of data; in the writing of the manuscript; or in the decision to publish the results.

References

1. Van de Watering, L. Red cell storage and prognosis. *Vox. Sang.* **2011**, *100*, 36–45. [CrossRef] [PubMed]
2. Hess, J.R. Red cell storage. *J. Proteomics* **2010**, *73*, 368–373. [CrossRef] [PubMed]
3. Chin-Yee, I.; Arya, N.; D'Almeida, M.S. The red cell storage lesion and its implication for transfusion. *Transfus. Sci.* **1997**, *18*, 447–458. [CrossRef]
4. Karkouti, K. From the Journal archives: the red blood cell storage lesion: Past, present, and future. *Can. J. Anaesth.* **2014**, *61*, 583–586. [CrossRef] [PubMed]
5. Hess, J.R. Red cell changes during storage. *Transfus. Apher. Sci.* **2010**, *43*, 51–59. [CrossRef] [PubMed]
6. Henkelman, S.; Dijkstra-Tiekstra, M.J.; De Wildt-Eggen, J.; Graaff, R.; Rakhorst, G.; Van Oeveren, W. Is red blood cell rheology preserved during routine blood bank storage? *Transfusion* **2010**, *50*, 941–948. [CrossRef] [PubMed]
7. Relevy, H.; Koshkaryev, A.; Manny, N.; Yedgar, S.; Barshtein, G. Blood banking induced alteration of red blood cell flow properties. *Transfusion* **2008**, *48*, 136–146. [CrossRef] [PubMed]
8. D'Alessandro, A.; Kriebardis, A.G.; Rinalducci, S.; Antonelou, M.H.; Hansen, K.C.; Papassideri, I.S.; Zolla, L. An update on red blood cell storage lesions, as gleaned through biochemistry and omics technologies. *Transfusion* **2015**, *55*, 205–219. [CrossRef] [PubMed]
9. Wither, M.; Dzieciatkowska, M.; Nemkov, T.; Strop, P.; D'Alessandro, A.; Hansen, K.C. Hemoglobin oxidation at functional amino acid residues during routine storage of red blood cells. *Transfusion* **2016**, *56*, 421–426. [CrossRef] [PubMed]
10. Rinalducci, S.; D'Amici, G.M.; Blasi, B.; Vaglio, S.; Grazzini, G.; Zolla, L. Peroxiredoxin-2 as a candidate biomarker to test oxidative stress levels of stored red blood cells under blood bank conditions. *Transfusion* **2011**, *51*, 1439–1449. [CrossRef] [PubMed]
11. Kriebardis, A.G.; Antonelou, M.H.; Stamoulis, K.E.; Economou-Petersen, E.; Margaritis, L.H.; Papassideri, I.S. Progressive oxidation of cytoskeletal proteins and accumulation of denatured hemoglobin in stored red cells. *J. Cell. Mol. Med.* **2007**, *11*, 148–155. [CrossRef] [PubMed]
12. Kriebardis, A.G.; Antonelou, M.H.; Stamoulis, K.E.; Economou-Petersen, E.; Margaritis, L.H.; Papassideri, I.S. Membrane protein carbonylation in non-leukodepleted CPDA-preserved red blood cells. *Blood. Cells. Mol. Dis.* **2006**, *36*, 279–282. [CrossRef] [PubMed]
13. Zhang, Z.Z.; Lee, E.E.; Sudderth, J.; Yue, Y.; Zia, A.; Glass, D.; Deberardinis, R.J.; Wang, R.C. Glutathione depletion, pentose phosphate pathway activation, and hemolysis in erythrocytes protecting cancer cells from vitamin C-induced oxidative stress. *J. Biol. Chem.* **2016**, *291*, 22861–22867. [CrossRef] [PubMed]

14. Yoshida, T.; Blair, A.; D'alessandro, A.; Nemkov, T.; Dioguardi, M.; Silliman, C.C.; Dunham, A. Enhancing uniformity and overall quality of red cell concentrate with anaerobic storage. *Blood. Transfus.* **2017**, *15*, 172–181. [CrossRef] [PubMed]

15. Chang, A.L.; Kim, Y.; Seitz, A.P.; Schuster, R.M.; Pritts, T.A. pH modulation ameliorates the red blood cell storage lesion in a murine model of transfusion. *J. Surg. Res.* **2017**, *212*, 54–59. [CrossRef] [PubMed]

16. Hess, J.R.; Hill, H.R.; Oliver, C.K.; Lippert, L.E.; Greenwalt, T.J. Alkaline CPD and the preservation of RBC 2,3-DPG. *Transfusion* **2002**, *42*, 747–752. [CrossRef] [PubMed]

17. Parthasarathi, K.; Lipowsky, H.H. Capillary recruitment in response to tissue hypoxia and its dependence on red blood cell deformability. *Am. J. Physiol.* **1999**, *277*, H2145–H2157. [PubMed]

18. Silliman, C.C.; Voelkel, N.F.; Allard, J.D.; Elzi, D.J.; Tuder, R.M.; Johnson, J.L.; Ambruso, D.R. Plasma and lipids from stored packed red blood cells cause acute lung injury in an animal model. *J. Clin. Investig.* **1998**, *101*, 1458–1467. [CrossRef] [PubMed]

19. Raval, J.S.; Fontes, J.; Banerjee, U.; Yazer, M.H.; Mank, E.; Palmer, A.F. Ascorbic acid improves membrane fragility and decreases haemolysis during red blood cell storage. *Transfus. Med.* **2013**, *23*, 87–93. [CrossRef] [PubMed]

20. Tzounakas, V.L.; Kriebardis, A.G.; Georgatzakou, H.T.; Foudoulaki-Paparizos, L.E.; Dzieciatkowska, M.; Wither, M.J.; Nemkov, T.; Hansen, K.C.; Papassideri, I.S.; D'Alessandro, A.; et al. Glucose 6-phosphate dehydrogenase deficient subjects may be better "storers" than donors of red blood cells. *Free. Radic. Biol. Med.* **2016**, *96*, 152–165. [CrossRef] [PubMed]

21. Stowell, S.R.; Smith, N.H.; Zimring, J.C.; Fu, X.; Palmer, A.F.; Fontes, J.; Banerjee, U.; Yazer, M.H. Addition of ascorbic acid solution to stored murine red blood cells increases posttransfusion recovery and decreases microparticles and alloimmunization. *Transfusion* **2013**, *53*, 2248–2257. [CrossRef] [PubMed]

22. Czubak, K.; Antosik, A.; Cichon, N.; Zbikowska, H.M. Vitamin C and Trolox decrease oxidative stress and hemolysis in cold-stored human red blood cells. *Redox. Rep.* **2017**, *11*, 1–6. [CrossRef] [PubMed]

23. Mohammed, B.M.; Fisher, B.J.; Huynh, Q.K.; Wijesinghe, D.S.; Chalfant, C.E.; Brophy, D.F.; Fowler, A.A., 3rd; Natarajan, R. Resolution of sterile inflammation: Role for vitamin C. *Mediat. Inflamm.* **2014**, *2014*, 173403. [CrossRef] [PubMed]

24. Bessis, M. Red cell shapes. An illustrated classification and its rationale. *Nouv. Rev. Fr. Hematol.* **1972**, *12*, 721–745.

25. Berezina, T.L.; Zaets, S.B.; Morgan, C.; Spillert, C.R.; Kamiyama, M.; Spolarics, Z.; Deitch, E.A.; Machiedo, G.W. Influence of storage on red blood cell rheological properties. *J. Surg. Res.* **2002**, *102*, 6–12. [CrossRef] [PubMed]

26. Lion, N.; Crettaz, D.; Rubin, O.; Tissot, J.D. Stored red blood cells: A changing universe waiting for its map(s). *J. Proteomics* **2010**, *73*, 374–385. [CrossRef] [PubMed]

27. Janz, D.R.; Zhao, Z.; Koyama, T.; May, A.K.; Bernard, G.R.; Bastarache, J.A.; Young, P.P.; Ware, L.B. Storage duration of red blood cells and risk of developing acute lung injury in patients with sepsis. *Am. J. Respir. Crit. Care Med.* **2013**, *3*, 33. [CrossRef]

28. May, J.M.; Qu, Z.C.; Qiao, H.; Koury, M.J. Maturational loss of the vitamin C transporter in erythrocytes. *Biochem. Biophys. Res. Commun.* **2007**, *360*, 295–298. [CrossRef] [PubMed]

29. Bianchi, J.; Rose, R.C. Glucose-independent transport of dehydroascorbic acid in human erythrocytes. *Proc. Soc. Exp. Biol. Med.* **1986**, *181*, 333–337. [CrossRef] [PubMed]

30. Mendiratta, S.; Qu, Z.C.; May, J.M. Erythrocyte ascorbate recycling: Antioxidant effects in blood. *Free Radic. Biol. Med.* **1998**, *24*, 789–797. [CrossRef]

31. Dawson, R.B. Blood storage XXV: Ascorbic acid (vitamin C) and dihydroxyacetone (DHA) maintenance of 2,3-DPG for six weeks in CPD-adenine. *Transfusion* **1977**, *17*, 248–254. [CrossRef] [PubMed]

32. Carr, A.; Frei, B. Does vitamin C act as a pro-oxidant under physiological conditions? *FASEB J.* **1999**, *13*, 1007–1024. [PubMed]

33. Herbert, V.; Shaw, S.; Jayatilleke, E. Vitamin C-driven free radical generation from iron. *J. Nutr.* **1996**, *126*, 1213S–1220S. [PubMed]

34. Knight, J.A.; Voorhees, R.P.; Martin, L.; Anstall, H. Lipid peroxidation in stored red cells. *Transfusion* **1992**, *32*, 354–357. [CrossRef] [PubMed]

35. Washko, P.W.; Wang, Y.; Levine, M. Ascorbic acid recycling in human neutrophils. *J. Biol. Chem.* **1993**, *268*, 15531–15535. [PubMed]

36. Park, J.B. Reduction of dehydroascorbic acid by homocysteine. *Biochim. Biophys. Acta* **2001**, *1525*, 173–179. [CrossRef]

37. Tu, H.; Li, H.; Wang, Y.; Niyyati, M.; Wang, Y.; Leshin, J.; Levine, M. Low red blood cell vitamin C concentrations induce red blood cell fragility: A link to diabetes via glucose, glucose transporters, and dehydroascorbic acid. *EBioMedicine* **2015**, *2*, 1735–1750. [CrossRef] [PubMed]

Article

Appropriate Handling, Processing and Analysis of Blood Samples Is Essential to Avoid Oxidation of Vitamin C to Dehydroascorbic Acid

Juliet M. Pullar, Simone Bayer and Anitra C. Carr *

Department of Pathology and Biomedical Science, University of Otago, Christchurch, P.O. Box 4345, Christchurch 8140, New Zealand; juliet.pullar@otago.ac.nz (J.M.P.); simone.bayer@otago.ac.nz (S.B.)
* Correspondence: anitra.carr@otago.ac.nz; Tel.: +64-3364-1524

Received: 21 December 2017; Accepted: 9 February 2018; Published: 11 February 2018

Abstract: Vitamin C (ascorbate) is the major water-soluble antioxidant in plasma and its oxidation to dehydroascorbic acid (DHA) has been proposed as a marker of oxidative stress in vivo. However, controversy exists in the literature around the amount of DHA detected in blood samples collected from various patient cohorts. In this study, we report on DHA concentrations in a selection of different clinical cohorts (diabetes, pneumonia, cancer, and critically ill). All clinical samples were collected into EDTA anticoagulant tubes and processed at 4 °C prior to storage at −80 °C for subsequent analysis by HPLC with electrochemical detection. We also investigated the effects of different handling and processing conditions on short-term and long-term ascorbate and DHA stability in vitro and in whole blood and plasma samples. These conditions included metal chelation, anticoagulants (EDTA and heparin), and processing temperatures (ice, 4 °C and room temperature). Analysis of our clinical cohorts indicated very low to negligible DHA concentrations. Samples exhibiting haemolysis contained significantly higher concentrations of DHA. Metal chelation inhibited oxidation of vitamin C in vitro, confirming the involvement of contaminating metal ions. Although EDTA is an effective metal chelator, complexes with transition metal ions are still redox active, thus its use as an anticoagulant can facilitate metal ion-dependent oxidation of vitamin C in whole blood and plasma. Handling and processing blood samples on ice (or at 4 °C) delayed oxidation of vitamin C by a number of hours. A review of the literature regarding DHA concentrations in clinical cohorts highlighted the fact that studies using colourimetric or fluorometric assays reported significantly higher concentrations of DHA compared to those using HPLC with electrochemical detection. In conclusion, careful handling and processing of samples, combined with appropriate analysis, is crucial for accurate determination of ascorbate and DHA in clinical samples.

Keywords: vitamin C; ascorbate; dehydroascorbic acid; whole blood; plasma; EDTA; heparin; HPLC with electrochemical detection; pneumonia; cancer; diabetes; critically ill

---

1. Introduction

The role of vitamin C (ascorbate) in health and disease has been widely studied since its discovery in the 1930s [1]. Ascorbate is the primary water-soluble antioxidant in plasma and a major non-enzymatic antioxidant in tissues; it can protect many important biomolecules from oxidation as well as regenerate specific molecules, such as vitamin E and tetrahydrobiopterin [2,3]. It is ascorbate's enzyme cofactor activity, however, that is likely to be its most significant function in vivo, with the list of biosynthetic and regulatory enzymes for which ascorbate is an essential cofactor growing [4,5]. Ascorbate is a cofactor for the metalloenzymes required for collagen tertiary structure stabilisation, transport of fatty acids into mitochondria for the generation of metabolic energy, and synthesis of catecholamine neurotransmitters and various peptide hormones [6,7]. Recent research has shown a role for ascorbate as a cofactor for the dioxygenases involved in gene regulation through modulation of transcription factors and epigenetic marks [4,5]. These latter gene-regulatory functions of ascorbate are responsible for modifying numerous signaling and biosynthetic pathways in the body and thus could play a major role in its observed health and disease modifying effects.

The biological functions of ascorbate depend upon its ability to donate electrons [6]. Loss of one electron generates the ascorbyl radical intermediate, and loss of two electrons generates dehydroascorbate (DHA, which can also be formed via dismutation of the ascorbyl radical) [2]. Detection of DHA in blood samples collected from different patient cohorts has been proposed to be an indication of oxidative stress occurring in these patients [8–12]. However, other investigators have been unable to detect elevated concentrations of DHA in comparable patient cohorts or in smokers, who are known to be under enhanced oxidative stress [13–15]. Therefore, there appears to be controversy in the literature. However, ascorbate is extremely sensitive to oxidation and factors such as oxygen, high temperature, UV light, and iron and copper contamination accelerate this process [16]. DHA is also unstable at neutral pH, and is rapidly hydrolysed to 2,3-diketogulonic acid in solution [17]. As such, samples in which ascorbate is to be measured must be very carefully handled and processed to avoid any artefactual oxidation of ascorbate with associated formation of DHA, and/or DHA degradation. The accurate quantification of ascorbate, and by extension DHA, in biological samples is vital to determining its potential health effects.

The current gold standard for the measurement of ascorbate and DHA in biological samples is high performance liquid chromatography (HPLC) with electrochemical detection [18]. Older methods relied upon colourimetric or fluorometric detection of ascorbate or DHA (following complete oxidation of the sample), with variable sensitivities and specificities [18]. An acid or alcohol precipitation step is typically incorporated into sample processing to precipitate protein and stabilize the ascorbate and DHA: meta-phosphoric acid (MPA), trichloroacetic acid (TCA), perchloric acid (PCA) and methanol have all been used. They are often combined with a metal chelator such as ethylenediaminetetraacetic acid (EDTA) or diethylene-triaminepentaacetic acid (DTPA), to help prevent ex vivo ascorbate oxidation [18]. The type of anticoagulant used for blood sample collection may also impact on recovery of ascorbate, with EDTA and heparin tubes providing the most accurate measurement of both ascorbate and DHA [19].

Due to the apparent controversy in the literature surrounding concentrations of DHA in clinical samples, we herein report on the measurement of DHA in a selection of different clinical cohorts (diabetes, pneumonia, cancer, and critically ill) using HPLC with electrochemical detection. We have also determined both the short-term and long-term stability of ascorbate in clinical samples treated under different conditions in order to better elucidate how blood samples should be handled and processed by investigators for accurate ascorbate and DHA assessment.

2. Materials and Methods

2.1. In Vitro Stability of Ascorbate and DHA

DHA is a dimer in the crystalline state and can be difficult to fully solubilise. Thus, the soluble DHA concentration was confirmed spectrophotometrically at 265 nm following reduction with 2.5 mmol/L dithiothreitol for 5 min at room temperature ($\varepsilon = 14500$ M^{-1} cm^{-1}). The in vitro stability of ascorbate (Sigma-Aldrich, Auckland, New Zealand) and DHA (Sigma-Aldrich, Auckland, New Zealand) in phosphate-buffered saline (PBS) was assessed spectrophotometrically. Fifty micromolar ascorbate or DHA was dissolved in PBS in the presence and absence of DTPA. Samples were removed at indicated time points and the absorbance at 265 nm determined ($\varepsilon = 14500$ M^{-1} cm^{-1}). Samples were blanked against buffer containing DTT when required.

2.2. Stability of Ascorbate in Blood and Plasma

Blood was obtained from healthy volunteers at the University of Otago, Christchurch. Ethical approval was granted from the Upper South A Regional Ethics Committee (URA/10/03/021) and all donors provided written informed consent. Blood samples were collected into EDTA or lithium heparin anticoagulant tubes as indicated. To prepare plasma, samples were centrifuged at $1000\times g$ for 10 min at 4 °C. Plasma was used immediately or alternatively stored at −80 °C until required.

The short-term stability of ascorbate and DHA in whole blood and isolated plasma was determined following collection into sterile EDTA or heparin vacutainer tubes in a sufficient amount for 24 h experiments. Following sterile distribution, tubes of blood or plasma were incubated in the dark, on ice, at 4 °C or at room temperature, and at indicated time points a sample was removed for processing for ascorbate HPLC analysis.

2.3. Long-Term Stability of Ascorbate and DHA

The long-term stability of ascorbate and DHA at −80 °C was assessed. Triplicates of standard DHA (~50 μmol/L) were prepared in PBS, aliquoted into cryovials and stored at −80 °C. Similarly triplicates of plasma and PCA-extracts of plasma were aliquoted into cryovials and stored at −80 °C. Triplicates of each sample were removed monthly for one year and ascorbate analysis undertaken. Analysis of the DHA samples required reduction with tris(2-carboxyethyl)phosphine hydrochloride (TCEP) prior to HPLC analysis.

2.4. Detection of DHA in Clinical Samples

Clinical samples which had been collected for vitamin C observational and interventional studies and which had had ascorbate analysis carried out in the presence and absence of the reducing agent TCEP were included in the current analysis. These studies included samples that were processed for ascorbate HPLC analysis prior to storage: prediabetes cohort (n = 24, ACTRN12616000858493), colorectal cancer cohort (n = 30, ACTRN12615001277538), community acquired pneumonia cohort (n = 24) and haematological cancer cohort (n = 6, HDEC16STH235), and critically ill surgical cohort (n = 17) and septic shock cohort (n = 24) [20]; or samples that were stored as frozen plasma: community cohort (n = 20) [21], and normal glucose control cohort (n = 32), prediabetes cohort (n = 24), and type 2 diabetes mellitus cohort (n = 30) [22]. All clinical samples were collected into EDTA anticoagulant tubes, processed at 4 °C, and stored at −80 °C prior to HPLC analysis.

2.5. Sample Preparation for HPLC Analysis

An aliquot of fresh (or previously frozen) EDTA or heparin plasma was treated with an equal volume of ice-cold 0.54 mol/L HPLC-grade perchloric acid (PCA) solution containing 100 μmol/L of the metal chelator diethylenetriaminepentaacetic acid (DTPA) to precipitate proteins and stabilize the ascorbate [23]. Samples were vortexed, incubated on ice for a few minutes, then centrifuged at 4 °C to remove precipitate. Samples were further diluted with an equal volume of ice cold 77 mmol/L PCA solution containing 100 μmol/L DTPA prior to HPLC analysis.

TCEP is a strong reducing agent that has recently been shown to reduce DHA to ascorbate at low pH [24,25]. To confirm that TCEP would reduce DHA using our extraction conditions with PCA, concentration and time-dependent experiments were undertaken with standard DHA. We found that TCEP was an effective reductant using the following conditions: 32 mmol/L for 3 h on ice (data not shown). Thus, a 100 μL aliquot of the PCA-supernatant was treated with 10 μL of TCEP (100 mg/mL stock) for 3 h at 4 °C. Samples were further diluted with an equal volume of 77 mmol/L PCA solution containing 100 μmol/L DTPA prior to HPLC analysis. DHA is calculated as the difference between total vitamin C (with TCEP) and ascorbate values (without TCEP).

2.6. Ascorbate HPLC Analysis

The ascorbate content of the samples was determined by reverse phase HPLC with electrochemical detection [23]. Samples were separated on a Synergi 4 μ Hydro-RP 80 A column 150 × 4.6 mm (Phenomenex NZ Ltd., Auckland, New Zealand) using an Ultimate 3000 HPLC unit with an Ultimate 3000 ECD-3000RS electrochemical detector and a Model 6011RS coulometric cell. The settings were as follows: +250 mV analytical electrode potential; +300 mV guard cell potential; autosampler chilled to 4 °C and column temperature set to 30 °C. The mobile phase consisted of 80 mmol/L sodium acetate buffer, pH 4.8, containing DTPA (0.54 mmol/L) with the ion-pair reagent n-octylamine (1 μmol/L) added just prior to use. The buffer was delivered isocratically at a flow rate of 1.2 mL/min and 20 μL of sample injected. A standard curve of sodium-L-ascorbate, standardised spectrophotometrically at 245 nm ($\varepsilon = 9860 \text{ M}^{-1} \text{ cm}^{-1}$), was freshly prepared for each HPLC run in 77 mmol/L HPLC-grade PCA containing DTPA (100 μmol/L). Plasma ascorbate concentration is expressed as μmol/L.

Control experiments were carried out to determine the stability of PCA/DTPA extracts in the HPLC autosampler chilled to 4 °C ($n = 4–5$). Injections were performed once per hour over 24 h. In the absence of the reducing agent TCEP, PCA extracts of both EDTA and heparin plasma were stable for up to 16 h (with a >10% loss observed from this time, data not shown). In comparison, ascorbate was stable for the entire 24 h tested in the presence of TCEP (data not shown).

2.7. Statistical Analyses

Data is represented as the mean ± SD for the clinical samples and mean ± SEM for the in vitro experiments, with the *p* value for statistical significance set at 0.05. Linear regression analysis was used to determine stability of ascorbate and DHA in long-term storage experiments. For short-term stability experiments, statistical difference from time zero was determined by repeated measures one-way analysis of variance with Dunnett's test for multiple comparison. Statistical analyses were carried out using GraphPad Prism version 7.03 (La Jolla, CA, USA).

3. Results

3.1. Detection of DHA in Clinical Samples

Analysis of clinical samples in the presence and absence of the reducing agent TCEP allowed us to quantify the amount of DHA present through subtraction of ascorbate (without TCEP) from total vitamin C (with TCEP). Table 1 shows concentrations of DHA in plasma samples from various patient cohorts that were processed and stabilized with PCA and DTPA prior to storage at −80 °C. DHA concentrations were very low to negligible. Only samples that exhibited haemolysis contained appreciable amounts of DHA (Table 1), likely due to ex vivo release of iron from haemoglobin during sample processing (protein precipitation step) and subsequent metal-ion dependent oxidation of ascorbate (see in vitro experiments below).

Table 1. Dehydroascorbic acid (DHA) concentrations in clinical samples.

Cohort	Description (n)	DHA (μmol/L) [1]
Community	Prediabetes (22)	2.8 ± 4.2
Infection	Pneumonia (20)	1.2 ± 1.2
Cancer	Colorectal (27)	2.5 ± 3.3
	Haematological (6)	1.0 ± 0.4
Critically ill	Septic shock (24)	0.1 ± 2.2
	Surgical (15)	0.0 ± 1.5
Excluded [2]	Heamolysed (9)	17 ± 11

[1] Values represent mean ± SD. [2] Haemolysed samples were excluded and presented separately: 2 from prediabetes, 2 from pneumonia, 3 from colorectal cancer, and 2 from surgical cohorts.

In contrast, plasma samples that had been stored at −80 °C prior to processing for ascorbate HPLC analysis exhibited ~10-fold higher concentrations of DHA than samples processed and stabilized with

PCA and DTPA prior to storage (Table 2). Long-term storage experiments indicated that DHA was relatively stable at $-80\,°C$ for up to 1 year, with <8% loss over this time ($p = 0.097$). This indicates that if any ascorbate is oxidized to DHA in plasma during sample processing and/or storage, due to the relative stability of DHA at $-80\,°C$, it should be possible to recover the oxidized ascorbate using TCEP prior to HPLC analysis. As our other long-term storage experiments indicated that ascorbate is stable at $-80\,°C$ for at least 1 year, both as plasma (94% remaining at 52 weeks, $p = 0.562$) and PCA/DTPA extracts (101% remaining at 52 weeks, $p = 0.572$), this indicates that the DHA present in the stored plasma samples may have been generated during handling of the plasma samples prior to storage. In support of this, the amount of DHA in the clinical samples did not differ depending on how long the samples had been stored (from 4 months to 2 years, data not shown).

Table 2. DHA concentrations in stored plasma samples [1].

Cohort (n)	DHA (μmol/L)
Community (20)	24 ± 7
Normal glucose control (32)	10 ± 4
Prediabetes (24)	9 ± 5
Type 2 diabetes mellitus (30)	8 ± 6

[1] Values represent mean $\pm$ SD.

3.2. In Vitro Stability of Ascorbate and DHA

Ascorbate in buffer is stable for at least 6 h if kept on ice (or at 4 °C), with a small decrease in concentration observed at 24 h (Figure 1A). In contrast, ascorbate was much less stable when incubated at room temperature, with a statistically significant decrease from 2 h onwards ($p < 0.05$). The presence of the metal chelator DTPA prevented the loss of ascorbate, indicating that the loss observed at room temperature is due to metal ion-dependent oxidation of ascorbate by contaminating transition metal ions in the buffer. When the same experiment was undertaken with DHA in buffer, the concentration of DHA decreased over time at room temperature, with a half-life of about 1.5 h, but was relatively stable on ice for up to 6 h (Figure 1B). Addition of DTPA to the DHA solution did not inhibit its degradation, which occurs via hydrolysis rather than metal ion-dependent oxidation. As mentioned above, long term storage experiments indicated that DHA was relatively stable at $-80\,°C$ for at least 1 year, likely due to attenuation of hydrolysis reactions at this temperature.

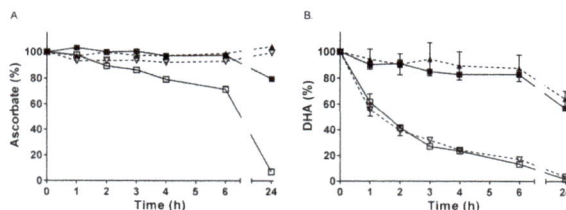

Figure 1. Stability of ascorbate and DHA in PBS. (**A**) Ascorbate (50 μmol/L) was incubated on ice (■,▲) or at room temperature (□,▽) in the absence (solid line) or presence (dashed line) of the metal chelator DTPA (100 μmol/L). (**B**) DHA (50 μmol/L) was incubated on ice (■,▲) or at room temperature (□,▽) in the absence (solid line) or presence (dashed line) of DTPA (100 μmol/L). Samples were analysed for ascorbate spectrophotometrically (following reduction of DHA with DTT) at the indicated time points. Data represent the mean $\pm$ SEM ($n = 3$). Repeated measure one-way ANOVA with Dunnett's test for multiple comparison showed no significant loss of ascorbate on ice over 6 h ($p = 0.697$), but a significant decrease at room temperature from 2 h (in the absence of DTPA, $p = 0.030$). A small but significant decrease in DHA on ice was observed from 3 h ($p = 0.042$) and a large decrease in DHA at room temperature was observed from 1 h ($p = 0.068$).

3.3. Stability of Ascorbate during Sample Collection and Processing

3.3.1. Stability in Whole Blood

Blood samples drawn into EDTA and heparin tubes were incubated for up to 24 h on ice or at room temperature (Figure 2). Ascorbate was stable for at least 6 h in both EDTA and heparin tubes when kept on ice (or at 4 °C). However, a time-dependent decrease in ascorbate was observed in EDTA blood when incubated at room temperature, with a significant decrease observed from 2 h onwards ($p < 0.05$). This loss could not be recovered via reduction of the samples with TCEP, likely due to the short half-life of DHA at room temperature.

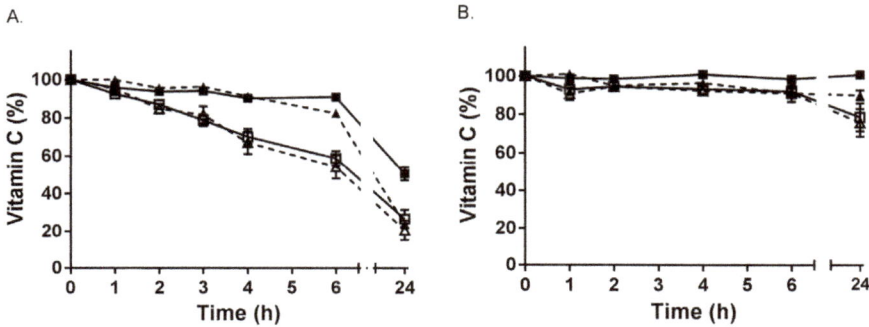

Figure 2. Stability of vitamin C in whole blood. (**A**) EDTA-blood was incubated on ice (■,▲) or at room temperature (□,△). (**B**) Heparin-blood was incubated on ice (■,▲) or at room temperature (□,△). Samples without TCEP reduction are shown with dashed lines, and those with TCEP using solid lines. Samples were extracted for ascorbate HPLC analysis at the indicated time points. Data represent the mean ± SEM ($n = 3$). Repeated measure one-way ANOVA with Dunnett's test for multiple comparison indicated no significant decrease in ascorbate in EDTA-blood on ice over 6 h ($p = 0.153$), but a significant decrease in ascorbate at room temperature from 2 h ($p = 0.046$). There was no significant decrease in ascorbate in heparin-blood on ice for 24 h ($p = 0.182$), or at room temperature over 6 h ($p = 0.163$).

3.3.2. Stability in Plasma

Plasma isolated from EDTA and heparin blood samples was incubated for up to 24 h on ice or at room temperature (Figure 3). Although ascorbate was relatively stable for the first few hours when EDTA plasma was kept on ice (or at 4 °C), a time-dependent decrease of ascorbate was observed at room temperature, with approximately 50% lost by 2 h. In comparison, ascorbate was more stable in heparin-plasma, with only a small drop in ascorbate observed over time when the sample was kept at room temperature (Figure 3B). Once again, the loss of ascorbate at room temperature could not be recovered following incubation with TCEP.

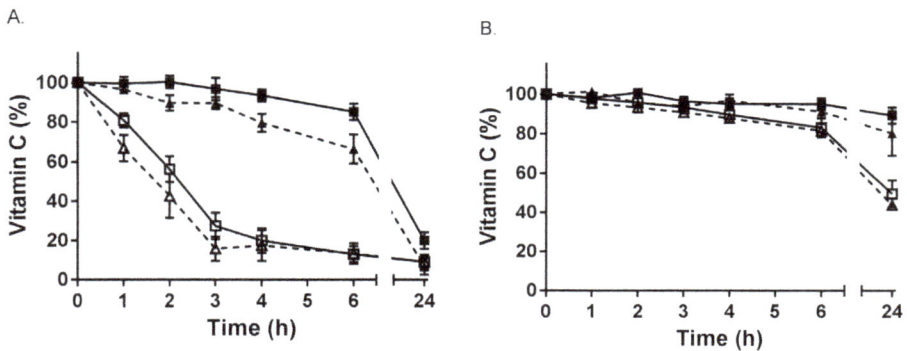

Figure 3. Stability of vitamin C in plasma. (**A**) EDTA-plasma was incubated on ice (■,▲) or at room temperature (□,△). (**B**) Heparin-plasma was incubated on ice (■,▲) or at room temperature (□,△). Samples without TCEP are shown with dashed lines, and those with TCEP using solid lines. Samples were extracted for ascorbate HPLC analysis at indicated time points. Data represent the mean ± SEM (*n* = 3–7). Repeated measure one-way ANOVA with Dunnett's test for multiple comparison showed a significant decrease in ascorbate in EDTA-plasma on ice from 4 h (in the absence of TCEP, *p* = 0.016), and at room temperature from 1 h (*p* = 0.026). There was no significant decrease in ascorbate in heparin-plasma on ice for 24 h (*p* = 0.544), or at room temperature over 6 h (*p* = 0.396).

4. Discussion

We have shown in a selection of different clinical cohorts that DHA can be detected in only very low micromolar amounts, if at all. All of the clinical samples were collected into EDTA anticoagulant tubes and were kept on ice/at 4 °C during sample handling and processing, and most were stabilized with PCA/DTPA prior to analysis or storage. The only samples that contained any appreciable amounts of DHA were those exhibiting haemolysis or those which had been stored as plasma prior to processing for ascorbate analysis. Koshiishi et al. [26] have previously shown that oxidation of ascorbate to DHA can occur following acidic deproteination due to catalysis by ferric ion released from haemoglobin and transferrin. When these investigators measured ascorbate directly, without acid deproteination, no DHA was observed, suggesting that DHA detected in acidified plasma samples is due to artefactual ascorbate oxidation [26]. Our in vitro experiments confirm that loss of vitamin C at physiological pH is due to the presence of contaminating metal ions, as indicated by complete inhibition of this loss with the metal chelator DTPA. This indicates that any samples containing haemolysis will require reduction to recover oxidized ascorbate prior to analysis. Although DHA is unstable at room temperature, exhibiting a half-life of ~1.5 h at physiological pH, we have shown that it is stable for several hours at 4 °C, and also exhibits relatively good long-term stability at −80 °C at physiological pH. Others have shown that, like ascorbate, DHA also has good long-term stability in acidified solutions [27,28]. This indicates that if any ascorbate should become oxidized to DHA during sample handling and processing, due to the relative stability of DHA at −80 °C, it should be possible to recover the oxidized ascorbate using an appropriate reducing agent.

Our short-term stability experiments in whole blood and plasma samples indicated that although ascorbate was stable for at least several hours at 4 °C, there was clear loss of ascorbate in EDTA whole blood if left at room temperature for 2 h and rapid loss, even within 1 h, in EDTA plasma. EDTA is routinely used as an anticoagulant for clinical samples as it chelates calcium ions which are required for clotting enzymes [29]. EDTA also chelates transition metal ions, such as iron and copper, which can cause oxidation of ascorbate. Therefore, EDTA is often added as a chelating agent to samples requiring measurement of ascorbate. However, EDTA chelates of iron and copper are not redox inactive, in other words, they are still capable of oxidizing ascorbate [30,31]. Although, addition of EDTA can help stabilize ascorbate in acidic solutions at 4 °C [18,32], it is clearly unable to stabilize ascorbate at physiological pH, particularly at room temperature [33]. Therefore, if ascorbate is to be measured in EDTA blood samples, it is essential that these are kept cold at all times during handling, processing and analysis of the samples.

A number of studies have previously emphasised the importance of the collection and processing method for maintaining ascorbate stability in blood samples, with temperature and pH being the most important factors, particularly with respect to the anticoagulant used and perhaps also the acid precipitant, although the latter is less clear [33–40]. Our review of the literature of DHA concentrations reported in clinical cohorts (summarized in Table 3) has also highlighted that high concentrations of DHA are observed in all studies (except one [13]) that have used colourimetric or flourometric methods for detecting DHA [8–12]. In comparison, DHA levels in clinical cohorts are much lower when HPLC with electrochemical detection is used [14,15], regardless of the anticoagulant [19,41] or the precipitating agent used [14,26]. This is likely due to the lack of specificity of the colourimetric/fluorometric methods, which are prone to interference by numerous different compounds in the blood, including transition metal ions [18]. It is noteworthy that depleted vitamin C concentrations, which are typically observed during severe infection, are associated with enhanced haemolysis [42], thus potentially contributing interfering metal ions to the colourimetric/fluorometric assays. Overall, this suggests that DHA may not be present in circulation and detection is due to assay conditions resulting in ex vivo ascorbate oxidation [41]. Furthermore, since DHA is rapidly taken up by cells via their glucose transporters (GLUTs), followed by intracellular reduction to ascorbate via enzymatic and non-enzymatic means [43,44], it would be very surprising indeed to detect appreciable amounts of DHA in the circulation.

Table 3. DHA concentrations in published clinical studies.

Cohort (n)	DHA (μmol/L) [1]	Anticoagulant, Deproteinization, and Detection	Ref.
Controls (28)	3 ± 1		
Pneumonia–died (7)	39 ± 2	Oxalate	
Pneumonia–survived (15)	23 ± 1	SSA	[8]
Convalescent (13)	9 ± 1	DCPIP	
Controls (10)	5 ± 2	Heparin	
Viral hepatitis (26)	29 ± 8	TCA	[9]
Liver carcinoma (11)	32 ± 5	DNPH	
Controls (20)	12 ± 4	Serum	
Rheumatoid arthritis (13)	22 ± 9	MPA/EDTA	[10]
		PDA/HPLC	
Controls (37)	0	Heparin	
Diabetic–male (25)	12 ± 2	MPA	[11]
Diabetic–female (12)	12 ± 2	DNPH	
Controls (20)	2.0 ± 1.5	EDTA	
Diabetic (27)	3.3 ± 3.0	MPA/oxalate	[13]
		DNPH	
Controls (34)	2.3 (−2.9–5.8)		
Critically ill (62)	1.4 (−0.8–2.9)	Heparin	
Diabetic (24)	2.8 (0.0–6.3)	MPA	[14]
Gastritis (21)	2.3 (0.6–2.9)	HPLC	
Controls (124)	0.1 ± 2.4	MPA	
Smokers (82)	0.8 ± 2.3	HPLC-ECD	[15]

[1] Values represent mean ± SD or mean (and range). Abbreviations: SSA, sulfosalicylic acid; TCA, trichloroacetic acid; MPA, metaphosphoric acid; EDTA, ethylenediaminetetraacetic acid; DCPIP, 2,6-dichlorophenol indophenol; DNPH, 2,4-dinitrophenylhydrazine; PDA, 1,2-phenylenediamine; ECD, electrochemical detection.

Despite the plethora of data on vitamin C stability under various assay conditions, clinical studies have been published utilizing less than ideal handling and processing conditions (e.g., EDTA blood stored at room temperature for up to 5 h followed by colourimetric detection), which results in low concentrations of ascorbate detected (i.e., 32 µmol/L) and a high percentage of deficiency (6% with plasma concentrations <11 µmol/L) [45]. This is in comparison to other age-matched studies where EDTA blood was handled on ice or at 4 °C and analyses were carried out by HPLC with electrochemical detection [21,46]. Similarly, the low serum vitamin C concentrations reported in a cohort of young Canadians (i.e., 24–30 µmol/L with 14% deficiency), despite saturating intakes (i.e., 228–248 mg/d) [47], has been questioned by Hoffer, due to likely issues with sample handling, processing and storage [48]. Thus, due to the numerous different factors that can affect ascorbate stability in clinical samples during handling, processing, storage and analysis, it is vital to assess ascorbate stability under the specific conditions being used in each study.

5. Conclusions

Our study and accompanying literature review have highlighted a number of important points with respect to accurate analysis of ascorbate and DHA in clinical study samples:

- Plasma stored at −80 °C prior to processing for ascorbate analysis contains variable amounts of DHA, requiring treatment with a reducing agent prior to ascorbate analysis (DHA is relatively stable for at least a year at −80 °C at physiological pH).
- Haemolysis facilitates oxidation of ascorbate, likely due to the release of catalytic iron from haemoglobin following acid precipitation, requiring reduction of samples prior to ascorbate analysis (DHA is stable for at least five years at −80 °C under acidic conditions).
- EDTA anticoagulant samples need to be kept cold at all times during handling, processing and analysis as EDTA-chelated iron is redox active at physiological pH and can facilitate ascorbate oxidation via redox cycling.
- Colourimetric/fluorometric ascorbate assays appear to generate high concentrations of DHA via artefactual ex vivo oxidation of ascorbate, regardless of the anticoagulant or deproteinization method used. In comparison, analysis using HPLC with electrochemical detection does not detect appreciable DHA concentrations in clinical samples.

Acknowledgments: Thanks to Renee Wilson and Margreet Vissers for access to the diabetes and colorectal cancer cohort samples, and thanks to all the clinical staff who recruited and collected the patient samples. We also thank Susannah Dunham (S.D.) for assistance with sample processing. S.D. was supported by a University of Otago, Christchurch, Summer Studentship (from Canterbury Health Laboratories). S.B. and J.P. were supported by a grant from the Canterbury Medical Research Foundation (PRO15/06, awarded to A.C.). A.C. is supported by a Health Research Council of New Zealand Sir Charles Hercus Health Research Fellowship (#16/037).

Author Contributions: A.C. conceived the study and managed the clinical cohort analyses, A.C. and J.P. designed the in vitro experiments, J.P. and S.B. performed the experiments, A.C. and S.B. analysed the clinical samples, A.C. and J.P. analysed the data and wrote the paper.

Conflicts of Interest: The authors declare no conflict of interest.

References

1. Svirbely, J.L.; Szent-Gyorgyi, A. The chemical nature of vitamin C. *Biochem. J.* **1933**, *27*, 279–285. [CrossRef] [PubMed]
2. Carr, A.; Frei, B. Does vitamin C act as a pro-oxidant under physiological conditions? *FASEB J.* **1999**, *13*, 1007–1024. [CrossRef] [PubMed]
3. Carr, A.C.; Zhu, B.Z.; Frei, B. Potential antiatherogenic mechanisms of ascorbate (vitamin C) and alpha-tocopherol (vitamin E). *Circ. Res.* **2000**, *87*, 349–354. [CrossRef] [PubMed]
4. Vissers, M.C.; Kuiper, C.; Dachs, G.U. Regulation of the 2-oxoglutarate-dependent dioxygenases and implications for cancer. *Biochem. Soc. Trans.* **2014**, *42*, 945–951. [CrossRef] [PubMed]
5. Young, J.I.; Zuchner, S.; Wang, G. Regulation of the epigenome by vitamin C. *Annu. Rev. Nutr.* **2015**, *35*, 545–564. [CrossRef] [PubMed]
6. Du, J.; Cullen, J.J.; Buettner, G.R. Ascorbic acid: Chemistry, biology and the treatment of cancer. *Biochim. Biophys. Acta* **2012**, *1826*, 443–457. [CrossRef] [PubMed]

7. Carr, A.C.; Shaw, G.M.; Fowler, A.A.; Natarajan, R. Ascorbate-dependent vasopressor synthesis: A rationale for vitamin C administration in severe sepsis and septic shock? *Crit. Care* **2015**, *19*, 418. [CrossRef] [PubMed]
8. Chakrabarti, B.; Banerjee, S. Dehydroascorbic acid level in blood of patients suffering from various infectious diseases. *Proc. Soc. Exp. Biol. Med.* **1955**, *88*, 581–583. [CrossRef] [PubMed]
9. Dubey, S.S.; Palodhi, G.R.; Jain, A.K. Ascorbic acid, dehydroascorbic acid and glutathione in liver disease. *Indian J. Physiol. Pharmacol.* **1987**, *31*, 279–283. [PubMed]
10. Lunec, J.; Blake, D.R. The determination of dehydroascorbic acid and ascorbic acid in the serum and synovial fluid of patients with rheumatoid arthritis (RA). *Free Radic. Res. Commun.* **1985**, *1*, 31–39. [CrossRef] [PubMed]
11. Chatterjee, I.B.; Banerjee, A. Estimation of dehydroascorbic acid in blood of diabetic patients. *Anal. Biochem.* **1979**, *98*, 368–374. [CrossRef]
12. Banerjee, A. Blood dehydroascorbic acid and diabetes mellitus in human beings. *Ann. Clin. Biochem.* **1982**, *19 Pt 2*, 65–70. [CrossRef] [PubMed]
13. Cox, B.D.; Whichelow, M.J. The measurement of dehydroascorbic acid and diketogulonic acid in normal and diabetic plasma. *Biochem. Med.* **1975**, *12*, 183–193. [CrossRef]
14. Schorah, C.J.; Downing, C.; Piripitsi, A.; Gallivan, L.; Al-Hazaa, A.H.; Sanderson, M.J.; Bodenham, A. Total vitamin C, ascorbic acid, and dehydroascorbic acid concentrations in plasma of critically ill patients. *Am. J. Clin. Nutr.* **1996**, *63*, 760–765. [CrossRef] [PubMed]
15. Lykkesfeldt, J.; Loft, S.; Nielsen, J.B.; Poulsen, H.E. Ascorbic acid and dehydroascorbic acid as biomarkers of oxidative stress caused by smoking. *Am. J. Clin. Nutr.* **1997**, *65*, 959–963. [CrossRef] [PubMed]
16. Buettner, G.R.; Jurkiewicz, B.A. Catalytic metals, ascorbate and free radicals: Combinations to avoid. *Radiat. Res.* **1996**, *145*, 532–541. [CrossRef] [PubMed]
17. Bode, A.M.; Cunningham, L.; Rose, R.C. Spontaneous decay of oxidized ascorbic acid (dehydro-L-ascorbic acid) evaluated by high-pressure liquid chromatography. *Clin. Chem.* **1990**, *36*, 1807–1809. [PubMed]
18. Washko, P.W.; Welch, R.W.; Dhariwal, K.R.; Wang, Y.; Levine, M. Ascorbic acid and dehydroascorbic acid analyses in biological samples. *Anal. Biochem.* **1992**, *204*, 1–14. [CrossRef]
19. Lykkesfeldt, J. Ascorbate and dehydroascorbic acid as biomarkers of oxidative stress: Validity of clinical data depends on vacutainer system used. *Nutr. Res.* **2012**, *32*, 66–69. [CrossRef] [PubMed]
20. Carr, A.C.; Rosengrave, P.C.; Bayer, S.; Chambers, S.; Mehrtens, J.; Shaw, G.M. Hypovitaminosis C and vitamin C deficiency in critically ill patients despite recommended enteral and parenteral intakes. *Crit. Care* **2017**, *21*, 300. [CrossRef] [PubMed]
21. Pearson, J.F.; Pullar, J.M.; Wilson, R.; Spittlehouse, J.K.; Vissers, M.C.M.; Skidmore, P.M.L.; Willis, J.; Cameron, V.A.; Carr, A.C. Vitamin C status correlates with markers of metabolic and cognitive health in 50-year-olds: Findings of the CHALICE cohort study. *Nutrients* **2017**, *9*, 831. [CrossRef] [PubMed]
22. Wilson, R.; Willis, J.; Gearry, R.; Skidmore, P.; Fleming, E.; Frampton, C.; Carr, A. Inadequate vitamin C status in prediabetes and type 2 diabetes mellitus: Associations with glycaemic control, obesity, and smoking. *Nutrients* **2017**, *9*, 997. [CrossRef] [PubMed]
23. Carr, A.C.; Pullar, J.M.; Moran, S.; Vissers, M.C.M. Bioavailability of vitamin C from kiwifruit in non-smoking males: Determination of 'healthy' and 'optimal' intakes. *J. Nutr. Sci.* **2012**, *1*, 1–9. [CrossRef] [PubMed]
24. Wechtersbach, L.; Cigic, B. Reduction of dehydroascorbic acid at low pH. *J. Biochem. Biophys. Methods* **2007**, *70*, 767–772. [CrossRef] [PubMed]
25. Sato, Y.; Uchiki, T.; Iwama, M.; Kishimoto, Y.; Takahashi, R.; Ishigami, A. Determination of dehydroascorbic acid in mouse tissues and plasma by using tris (2-carboxyethyl) phosphine hydrochloride as reductant in metaphosphoric acid/ethylenediaminetetraacetic acid solution. *Biol. Pharm. Bull.* **2010**, *33*, 364–369. [CrossRef] [PubMed]
26. Koshiishi, I.; Mamura, Y.; Liu, J.; Imanari, T. Evaluation of an acidic deproteinization for the measurement of ascorbate and dehydroascorbate in plasma samples. *Clin. Chem.* **1998**, *44*, 863–868. [PubMed]
27. Koshiishi, I.; Imanari, T. Measurement of ascorbate and dehydroascorbate contents in biological fluids. *Anal. Chem.* **1997**, *69*, 216–220. [CrossRef] [PubMed]
28. Lykkesfeldt, J. Ascorbate and dehydroascorbic acid as reliable biomarkers of oxidative stress: Analytical reproducibility and long-term stability of plasma samples subjected to acidic deproteinization. *Cancer Epidemiol. Prev. Biomark.* **2007**, *16*, 2513–2516. [CrossRef] [PubMed]

29. Banfi, G.; Salvagno, G.L.; Lippi, G. The role of ethylenediamine tetraacetic acid (EDTA) as in vitro anticoagulant for diagnostic purposes. *Clin. Chem. Lab. Med.* **2007**, *45*, 565–576. [CrossRef] [PubMed]

30. Burkitt, M.J.; Gilbert, B.C. Model studies of the iron-catalysed Haber-Weiss cycle and the ascorbate-driven Fenton reaction. *Free Radic. Res. Commun.* **1990**, *10*, 265–280. [CrossRef] [PubMed]

31. Buettner, G.R. Ascorbate oxidation: UV absorbance of ascorbate and ESR spectroscopy of the ascorbyl radical as assays for iron. *Free Radic. Res. Commun.* **1990**, *10*, 5–9. [CrossRef] [PubMed]

32. Salminen, I.; Alfthan, G. Plasma ascorbic acid preparation and storage for epidemiological studies using TCA precipitation. *Clin. Biochem.* **2008**, *41*, 723–727. [CrossRef] [PubMed]

33. Lee, W.; Davis, K.A.; Rettmer, R.L.; Labbe, R.F. Ascorbic acid status: Biochemical and clinical considerations. *Am. J. Clin. Nutr.* **1988**, *48*, 286–290. [CrossRef] [PubMed]

34. Karlsen, A.; Blomhoff, R.; Gundersen, T.E. Stability of whole blood and plasma ascorbic acid. *Eur. J. Clin. Nutr.* **2007**, *61*, 1233–1236. [CrossRef] [PubMed]

35. Cuerq, C.; Peretti, N.; Chikh, K.; Mialon, A.; Guillaumont, M.; Drai, J.; Blond, E. Overview of the in vitro stability of commonly measured vitamins and carotenoids in whole blood. *Ann. Clin. Biochem.* **2015**, *52 Pt 2*, 259–269. [CrossRef] [PubMed]

36. Chung, W.Y.; Chung, J.K.; Szeto, Y.T.; Tomlinson, B.; Benzie, I.F. Plasma ascorbic acid: Measurement, stability and clinical utility revisited. *Clin. Biochem.* **2001**, *34*, 623–627. [CrossRef]

37. Liau, L.S.; Lee, B.L.; New, A.L.; Ong, C.N. Determination of plasma ascorbic acid by high-performance liquid chromatography with ultraviolet and electrochemical detection. *J. Chromatogr. B Biomed. Sci. Appl.* **1993**, *612*, 63–70. [CrossRef]

38. Ching, S.Y.; Prins, A.W.; Beilby, J.P. Stability of ascorbic acid in serum and plasma prior to analysis. *Ann. Clin. Biochem.* **2002**, *39*, 518–520. [CrossRef] [PubMed]

39. Bradley, D.W.; Emery, G.; Maynard, J.E. Vitamin C in plasma: A comparative study of the vitamin stabilized with trichloroacetic acid or metaphosphoric acid and the effects of storage at -70 degrees, -20 degrees, 4 degrees, and 25 degrees on the stabilized vitamin. *Clin. Chim. Acta* **1973**, *44*, 47–52. [CrossRef]

40. Kand'ar, R.; Zakova, P. Determination of ascorbic acid in human plasma with a view to stability using HPLC with UV detection. *J. Sep. Sci.* **2008**, *31*, 3503–3508. [CrossRef] [PubMed]

41. Dhariwal, K.R.; Hartzell, W.O.; Levine, M. Ascorbic acid and dehydroascorbic acid measurements in human plasma and serum. *Am. J. Clin. Nutr.* **1991**, *54*, 712–716. [CrossRef] [PubMed]

42. Tu, H.; Li, H.; Wang, Y.; Niyyati, M.; Wang, Y.; Leshin, J.; Levine, M. Low red blood cell vitamin C concentrations induce red blood cell fragility: A link to diabetes via glucose, glucose transporters, and dehydroascorbic acid. *EBioMedicine* **2015**, *2*, 1735–1750. [CrossRef] [PubMed]

43. Rumsey, S.C.; Kwon, O.; Xu, G.W.; Burant, C.F.; Simpson, I.; Levine, M. Glucose transporter isoforms GLUT1 and GLUT3 transport dehydroascorbic acid. *J. Biol. Chem.* **1997**, *272*, 18982–18989. [CrossRef] [PubMed]

44. Welch, R.W.; Wang, Y.; Crossman, A., Jr.; Park, J.B.; Kirk, K.L.; Levine, M. Accumulation of vitamin C (ascorbate) and its oxidized metabolite dehydroascorbic acid occurs by separate mechanisms. *J. Biol. Chem.* **1995**, *270*, 12584–12592. [CrossRef] [PubMed]

45. Johnston, C.S.; Thompson, L.L. Vitamin C status of an outpatient population. *J. Am. Coll. Nutr.* **1998**, *17*, 366–370. [CrossRef] [PubMed]

46. Gan, R.; Eintracht, S.; Hoffer, L.J. Vitamin C deficiency in a university teaching hospital. *J. Am. Coll. Nutr.* **2008**, *27*, 428–433. [CrossRef] [PubMed]

47. Cahill, L.; Corey, P.N.; El-Sohemy, A. Vitamin C deficiency in a population of young canadian adults. *Am. J. Epidemiol.* **2009**, *170*, 464–471. [CrossRef] [PubMed]

48. Hoffer, L.J. Re: "Vitamin C deficiency in a population of young canadian adults". *Am. J. Epidemiol.* **2010**, *171*, 387. [CrossRef] [PubMed]

antioxidants

MDPI

Article

High Vitamin C Status Is Associated with Elevated Mood in Male Tertiary Students

Juliet M. Pullar *, Anitra C. Carr, Stephanie M. Bozonet and Margreet C. M. Vissers

Centre for Free Radical Research, Department of Pathology and Biomedical Science, University of Otago, Christchurch, P.O. Box 4345, Christchurch 8140, New Zealand; anitra.carr@otago.ac.nz (A.C.C.); stephanie.bozonet@otago.ac.nz (S.M.B.); margreet.vissers@otago.ac.nz (M.C.M.V.)
* Correspondence: juliet.pullar@otago.ac.nz; Tel.: +64-3365-1559

Received: 13 June 2018; Accepted: 12 July 2018; Published: 16 July 2018

Abstract: Micronutrient status is thought to impact on psychological mood due to the role of nutrients in brain structure and function. The aim of the current study was to investigate the association of vitamin C status with mood state in a sample of male tertiary students. We measured fasting plasma vitamin C levels as an indicator of vitamin C status, and subjective mood was determined using the Profile of Mood States (POMS) questionnaire. One hundred and thirty-nine male students aged 18 to 35 years were recruited from local tertiary institutes in Christchurch, New Zealand. The average plasma vitamin C concentration was 58.2 ± 18.6 (SD) µmol/L and the average total mood disturbance score was 25.5 ± 26.6 (possible score -32 to 200 measuring low to high mood disturbance, respectively). Plasma vitamin C concentration was inversely correlated with total mood disturbance as assessed by POMS ($r = -0.181$, $p < 0.05$). Examination of the individual POMS subscales also showed inverse associations of vitamin C status with depression, confusion, and anger. These findings suggest that high vitamin C status may be associated with improved overall mood in young adult males.

Keywords: vitamin C; ascorbate; plasma; mood; total mood disturbance; POMS

1. Introduction

Evidence is accumulating that increased consumption of fruit and vegetables is associated with enhanced mood and psychological well-being [1–5]. While it is possible that fruit and vegetable intake is simply a marker of a "healthier" lifestyle, fruit and vegetables are rich in micronutrients, and there are a number of these that may contribute to an effect on mood [6,7]. Fatigue and depression are known to closely precede the physical symptoms of scurvy—a disease caused by vitamin C deficiency [8,9]—suggesting that vitamin C may also be a moderator of mood. Although known for its antioxidant properties, vitamin C (ascorbate) is also a cofactor for a family of biosynthetic and regulatory enzymes with important functions throughout the body. Critically, it is required for the synthesis of the monoamine neurotransmitters dopamine, noradrenaline, and possibly serotonin [10], deficiencies and dysregulation of which have been hypothesised to contribute to depression [11]. Vitamin C is also a cofactor for enzymes involved in the synthesis of carnitine, which is required for the generation of metabolic energy and has been implicated in the fatigue and lethargy associated with scurvy [10,12]. Furthermore, vitamin C regulates the epigenome; it is a cofactor for enzymes involved in both DNA and histone demethylation [13,14]. Epigenetic modifications provide a mechanism by which environmental signals, such as stress, can alter gene expression and neural function and thereby affect behaviour, cognition, and mental health [15].

Vitamin C levels are tightly regulated throughout the body, and its distribution is generally thought to reflect a functional requirement [16]. Concentrations are highest in the brain and other neuroendocrine tissues such as the pituitary and adrenal glands [17]. Indeed, animal models have shown that the brain is the last organ to be depleted of vitamin C during prolonged deficiency, suggesting a vital role in this tissue [18,19].

Several observational studies have suggested a relationship between vitamin C status—typically measured by dietary intake—and mood [20–22]. These are further supported by a number of small intervention trials in which participants were supplemented with oral vitamin C [23–27]. Hoffer and co-workers found that supplementation with 1 g/day reduced mood disturbance and psychological distress in acutely hospitalised patients [24]. Similarly, a reduction in anxiety was observed in high school students given 500 mg/day of vitamin C compared to a placebo [23]. We have shown an improvement in subjective mood in a group of male tertiary students supplemented with two gold kiwifruit per day, a food source particularly high in vitamin C (~130 mg vitamin C per kiwifruit). A significant effect was observed in those individuals with higher total mood disturbance at baseline [25]. In addition to a decrease in total mood disturbance, a decrease in fatigue, an increase in vigour, and a trend towards a decrease in depression were demonstrated.

The aim of the current study was to investigate the association of vitamin C status with subjective mood. We measured fasting plasma vitamin C levels as an indicator of vitamin C status in a sample of male tertiary students, and subjective mood was determined using a Profile of Mood States (POMS) questionnaire. This test has been validated and shown to be reliable for assessing mood states [28].

2. Materials and Methods

2.1. Study Design

This cross-sectional survey was undertaken between April and September 2012. Sample size calculations based on our previous vitamin C studies [29,30] with a standard deviation of 16 μmol/L at 5% type I error indicated 140 participants were required. When sample size was calculated based on the POMS total mood disturbance (TMD) score, a standard deviation of 25 units (as per our previous study [25]) and precision of 5 units at 5% type I error indicated 96 participants were required. A total of 139 male tertiary-level students aged 18 to 35 years and residing in Christchurch, New Zealand at the time of the study were recruited. Recruitment was through verbal or visual/electronic advertisements at local tertiary institutes. The study received ethical approval from the Upper South B Regional Ethics Committee, Christchurch (ethics reference URB/11/12/048). Informed consent was obtained prior to interview and sampling. At the interview, the participants completed a mood questionnaire, a health and lifestyle questionnaire, and blood samples were taken. Participants' height and weight were recorded by the study interviewer, and the body mass index (BMI) was calculated (kg/m^2).

2.2. Vitamin C Analysis

Fasting peripheral blood was collected by venipuncture into 4 mL K$_3$-EDTA vacutainer tubes (Becton Dickinson, Auckland, New Zealand) and immediately placed on ice. Samples were centrifuged at 4 °C to separate plasma, and this was mixed with an equal volume of ice-cold 0.54 mol/L perchloric acid solution containing the metal chelator diethylene-triamine-penta-acetic acid (DTPA) [29]. After centrifugation, the deproteinated plasma samples were stored at -80 °C prior to analysis by HPLC with electrochemical detection, as described previously [29]. Plasma vitamin C concentration is expressed as μmol/L.

2.3. Analysis of Mood

The Profile of Mood States (POMS) questionnaire was used to determine the participants' mood during the previous week. Scores were calculated using a POMS standard scoring grid (Psychological Assessments, Australia). The form comprises 65 mood-related adjectives, which are rated on a 5-point Likert-type scale ranging from 0 (not at all) to 4 (extremely) and then categorised into six mood subscales: tension-anxiety, depression-dejection, anger-hostility, vigour-activity, fatigue-inertia, and confusion-bewilderment. A TMD score is calculated by adding the depression, fatigue, tension, anger, and confusion sub-scores and then subtracting the vigour score. TMD scores range from −32 to 200; a higher score indicates more severe mood disturbance [28].

2.4. Statistical Analysis

Data are represented as mean ± SD and 95% confidence intervals. Correlations were tested using Pearson's Correlation Coefficient with SPSS software (version 22, IBM Corp. Armonk, NY, USA) and differences between nonparametric independent samples used the Mann–Whitney U test; *p* values ≤ 0.05 were considered significant.

3. Results

One hundred and thirty-nine male students aged 18 to 35 years were recruited from local tertiary institutes in Christchurch, New Zealand. No exclusion criteria were applied. The baseline characteristics of the study participants are presented in Table 1. The range of plasma levels was 5 μmol/L to 101 μmol/L, with a mean fasting plasma vitamin C concentration of 58 μmol/L. These are normal fasting values. The majority of the cohort had adequate vitamin C concentrations of 50 μmol/L or greater (71%). Roughly one quarter of participants had inadequate vitamin C status of 23–50 μmol/L [31], 2% were marginal (i.e., 11–23 μmol/L), and 0.7% had actual vitamin C deficiency (i.e., <11 μmol/L). The average TMD score of the participants was 25.5, which is similar to values obtained for male college students in the United States [28].

Table 1. Characteristics of individuals who completed the study.

Participant Characteristics	Mean ± SD	95% CI	*n* (%)
Age (years)	21.2 ± 2.5	20.8, 21.6	-
Ethnicity	-	-	-
Maori	-	-	13 (9)
NZ European	-	-	106 (76)
Weight (kg)	81.6 ± 15.9	78.9, 84.3	-
Height (cm)	180 ± 7.3	178.8, 181.3	-
BMI (kg/m²)	25.1 ± 4.3	24.4, 25.8	-
Vitamin C (μmol/L)	58.2 ± 18.6	55.1, 61.3	-
Adequate	-	-	99 (71)
Inadequate	-	-	36 (26)
Marginal	-	-	3 (2)
Deficient	-	-	1 (0.7)
TMD score	25.5 ± 26.6	21.0, 30.0	-

TMD score was *n* = 138, otherwise data are for *n* = 139. Plasma vitamin C was classified as deficient <11 μmol/L, marginal 11–23 μmol/L, inadequate 23–50 μmol/L, or adequate >50 μmol/L. TMD, total mood disturbance; CI, confidence interval; NZ, New Zealand; BMI, body mass index.

We investigated the relationship between vitamin C status, as assessed by plasma vitamin C concentration, and subjective mood (Table 2). Plasma vitamin C concentration was inversely correlated with total mood disturbance, as assessed by POMS (r = −0.181, *p* < 0.05). Examination of the individual POMS subscales also showed inverse associations of vitamin C status with depression and anger.

Furthermore, when participants were split into two groups around either the average plasma vitamin C concentration of 58.2 μmol/L or the adequacy of their vitamin C status (50 μmol/L cut-off), higher total mood disturbance, as assessed by POMS, was associated with lower plasma vitamin C concentration (Figure 1A,B). When participants were divided around the mean plasma vitamin C concentration, median TMD scores were 25 (IQR 4-52) in the low vitamin C group and 17 (IQR 4-36) in the high vitamin C group. Similarly, participants split around the adequacy of vitamin C status had a median TMD score of 27 (IQR 13-53) in the inadequate group and 17.5 (IQR 3-37) in the adequate group. In addition, those with adequate vitamin C status had significantly lower POMS subscores for depression and confusion as compared to those with inadequate status (Table 3).

Table 2. Pearson linear correlations of plasma vitamin C with mood.

POMS Subscore	r	*p* Value
Total mood disturbance	−0.181	0.034
Depression	−0.192	0.024
Fatigue	−0.061	0.480
Tension	−0.098	0.255
Anger	−0.172	0.044
Vigour	0.100	0.245
Confusion	−0.148	0.084

Total mood disturbance is the sum of the depression, fatigue, tension, anger, and confusion subscores minus the vigour score; n = 138.

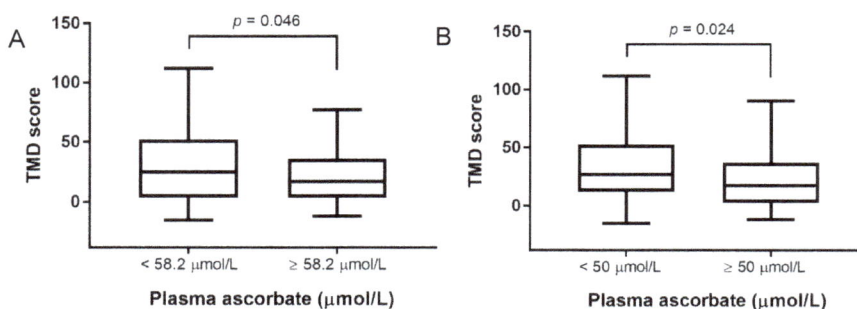

Figure 1. Relationship between total mood disturbance (TMD) score and plasma vitamin C concentration. (**A**) Participants were divided around the mean plasma vitamin C concentration of 58.2 μmol/L. (**B**) Participants were divided around adequacy of vitamin C status (a plasma concentration of 50 μmol/L indicates adequacy). Box plots show median TMD score with the 25th and 75th percentiles as boundaries; whiskers indicate the minimum and maximum of all the data. The TMD score was significantly different between the two groups for each graph (Mann–Whitney U test on ranks).

Table 3. Association of plasma vitamin C adequacy with Profile of Mood States (POMS) mood subscales.

POMS Subscore	*p* Value
Total mood disturbance	0.024
Depression	0.012
Fatigue	0.235
Tension	0.195
Anger	0.131
Vigour	0.453
Confusion	0.022

Participants were divided into two groups based on the adequacy of their vitamin C status (50 μmol/L cut-off). Differences in the TMD subscores were tested using the Mann–Whitney U test on ranks.

4. Discussion

The brain and central nervous system have a requirement for specific dietary nutrients [32,33]. Supplementation studies have shown an improvement in symptoms for certain mental health disorders with intake of nutrient formulations [34–38]; however, nutrients are also likely to be vital for normal psychological functioning and well-being in healthy individuals. The specific nutrients that are important for brain health are still being investigated. In the present study, we found a significant association between vitamin C status and current mood state in a sample of young adult males. Those individuals with the highest plasma vitamin C concentrations were more likely to have elevated mood.

Mood refers to a positive or negative emotional state of varying intensity that changes in response to life circumstances [39]. Mood is considered long-lasting in contrast to the more acutely experienced emotions. In our study, we used the POMS questionnaire to measure mood state during the previous week. As well as providing a total mood score, POMS gives five different measures of negative mood (depression, fatigue, tension, anger, and confusion) and a single measure of positive mood (vigour). In addition to the relationship observed with overall mood, we have shown significant inverse correlations of vitamin C status with the depression, anger, and confusion subscores in the young men studied. No relationship was observed with the positive mood state vigour despite our previous studies showing an increase in feelings of vigour with a food-based intervention that markedly elevated vitamin C levels [25] and despite emerging evidence for the association of dietary factors with positive well-being [1]. It should be noted that in the study cohort, there were only a few individuals with low vitamin C status of <23 µmol/L, meaning we were unable to investigate the mood state of this group. Rather, our results have shown that those with adequate vitamin C status (>50 µmol/L) tended to have an elevated mood.

One of the best-established functions of vitamin C is in the regulation of neurotransmitter biosynthesis, including that of catecholamines dopamine, norepinephrine, and epinephrine. Vitamin C acts as a cofactor for the enzyme dopamine β-hydroxylase, which converts dopamine to norepinephrine [40]. Animal models of vitamin C deficiency have shown decreased norepinephrine concentrations [41–43]. Furthermore, vitamin C can also recycle tetrahydrobiopterin, which is necessary for activation of tyrosine hydroxylase, the rate-limiting enzyme in catecholamine synthesis that synthesizes the dopamine precursor L-3,4-dihydroxyphenylalanine (L-DOPA) [44]. Similarly, tetrahydrobiopterin is a cofactor for tryptophan hydroxylase [45], the initial and rate-limiting enzyme in the synthesis of the neurotransmitter serotonin. There is also evidence emerging that vitamin C is involved in neuronal maturation and functioning [46]. Indeed, brain neurons contain some of the highest levels of vitamin C observed in any mammalian tissue [46]; glial ascorbate concentrations are much lower by comparison.

While the underlying pathophysiology of depression is not yet fully understood, these effects of vitamin C on neurochemistry may provide a mechanism by which it can affect this disorder. An early hypothesis suggested that deficiencies in dopamine, noradrenaline, and serotonin were responsible for major depressive symptoms [11], with some antidepressants elevating levels of these neurotransmitters in the central nervous system. However, it is now apparent that the molecular basis of depression is significantly more complex. Disturbances in dopamine, noradrenaline, and serotonin neurotransmission itself may contribute to the disorder. A more recent hypothesis suggests that low-grade inflammation and immune dysregulation, possibly as a result of psychosocial stressors, may trigger the development and persistence of depression [47]. For example, cytokines are known to induce depressive-type behaviours, and abnormal expressions of proinflammatory cytokines have been shown in patients with depression [48–50]. Oxidative stress markers are also elevated in patients with depression [51]. Vitamin C has a number of anti-inflammatory activities as well as being an excellent antioxidant and reducing agent, and it may be able to modulate some of these responses [12,52,53].

A limitation of the current study is that the data is cross-sectional and does not take into account potential confounders of the relationship between vitamin C status and mood, for example, socioeconomic status or other health behaviours. We did not determine the potential impact of any major recent life events that may affect mood in our cohort. Other unmeasured confounders may also have occurred simultaneously in our participants, such as deficiency in another micronutrient or a lower level of physical activity. Thus, we cannot definitively determine whether the relationship between vitamin C status is direct or, as influenced by the confounders above, indirect or parallel. Additionally, it may be that those with better mental health eat more fruits and vegetables causing higher vitamin C status, that is, higher vitamin C status is a consequence of better mood and mental health. In order to provide evidence of a direct relationship between plasma vitamin C status and mood, well-conducted randomized controlled trials are required. This will allow the direction of the relationship to be firmly established and will also allow the effect of any confounders to be eliminated as these should be evenly distributed between the two groups.

Levels of vitamin C in our cohort were generally higher than has been reported in other similar populations [54–56]. For example, a recent study in a cohort from Poland showed 7% of participants were marginally deficient in vitamin C [55], while a United States sample found 12–16% were marginally deficient [54]. Dietary information from our cohort indicated that there were a significant number of individuals who regularly used dietary supplements or consumed fruit juice containing vitamin C. It was our estimation that these dietary vitamin C sources contributed significantly to the high mean plasma status of our cohort. Apparent differences between the study populations may also reflect shortcomings in the sample handling and processing used in the studies described above, as inadequate processing can, and commonly does, increase the proportion of samples which are vitamin C deficient [57].

5. Conclusions

In conclusion, our findings suggest a possible relationship between vitamin C status and mood state in young adult male students in New Zealand. The current study is cross-sectional and further well-conducted intervention trials are required for proof of causality. There are a number of biological justifications for a positive effect of vitamin C on mood, particularly owing to its role in brain homeostasis and function.

Author Contributions: M.C.M.V. and J.M.P. conceived the study; J.M.P., A.C.C., and M.C.M.V. contributed to the study design; J.M.P. coordinated the study; J.M.P. and A.C.C. conducted the interviews; and S.M.B. measured vitamin C status. J.M.P. analysed the data and wrote the paper. All authors edited the paper.

Funding: This research received no external funding.

Acknowledgments: We express our gratitude to the young men who participated in this study. We acknowledge Jo Kepple for the use of Primorus Clinical Trials Unit, Susan Woods for the use of the Ara Institute of Canterbury Health Centre, and Joan Allardyce for the use of University of Canterbury Health Centre. We acknowledge John Pearson for statistical advice and Maria Webb for assistance with recruitment.

Conflicts of Interest: The authors declare no conflict of interest.

References

1. Blanchflower, D.G.; Oswald, A.J.; Stewart-Brown, S. Is Psychological Well-Being Linked to the Consumption of Fruit and Vegetables? *Soc. Indic. Res.* **2013**, *114*, 785–801. [CrossRef]
2. Conner, T.S.; Brookie, K.L.; Richardson, A.C.; Polak, M.A. On carrots and curiosity: Eating fruit and vegetables is associated with greater flourishing in daily life. *Br. J. Health Psychol.* **2015**, *20*, 413–427. [CrossRef] [PubMed]
3. Tsai, A.C.; Chang, T.L.; Chi, S.H. Frequent consumption of vegetables predicts lower risk of depression in older Taiwanese—Results of a prospective population-based study. *Public Health Nutr.* **2012**, *15*, 1087–1092. [CrossRef] [PubMed]
4. Bishwajit, G.; O'Leary, D.P.; Ghosh, S.; Sanni, Y.; Shangfeng, T.; Zhanchun, F. Association between depression and fruit and vegetable consumption among adults in South Asia. *BMC Psychiatry* **2017**, *17*, 15. [CrossRef] [PubMed]
5. Mujcic, R.; Oswald, J. Evolution of Well-Being and Happiness After Increases in Consumption of Fruit and Vegetables. *Am. J. Public Health* **2016**, *106*, 1504–1510. [CrossRef] [PubMed]

6. Rooney, C.; McKinley, M.C.; Woodside, J.V. The potential role of fruit and vegetables in aspects of psychological well-being: A review of the literature and future directions. *Proc. Nutr. Soc.* **2013**, *72*, 420–432. [CrossRef] [PubMed]

7. Kaplan, B.J.; Crawford, S.G.; Field, C.J.; Simpson, J.S. Vitamins, minerals, and mood. *Psychol. Bull.* **2007**, *133*, 747–760. [CrossRef] [PubMed]

8. Levine, M.; Conry-Cantilena, C.; Wang, Y.; Welch, R.W.; Washko, P.W.; Dhariwal, K.R.; Park, J.B.; Lazarev, A.; Graumlich, J.F.; King, J.; et al. Vitamin C pharmacokinetics in healthy volunteers: Evidence for a recommended dietary allowance. *Proc. Natl. Acad. Sci. USA* **1996**, *93*, 3704–3709. [CrossRef] [PubMed]

9. Crandon, J.H.; Lund, C.C.; Dill, D.B. Experimental human scurvy. *N. Engl. J. Med.* **1940**, *223*, 353–369. [CrossRef]

10. Englard, S.; Seifter, S. The biochemical functions of ascorbic acid. *Annu. Rev. Nutr.* **1986**, *6*, 365–406. [CrossRef] [PubMed]

11. Delgado, P.L. Depression: The case for a monoamine deficiency. *J. Clin. Psychiatry* **2000**, *61*, 7–11. [PubMed]

12. Du, J.; Cullen, J.J.; Buettner, G.R. Ascorbic acid: Chemistry, biology and the treatment of cancer. *Biochim. Biophys. Acta* **2012**, *1826*, 443–457. [CrossRef] [PubMed]

13. Minor, E.A.; Court, B.L.; Young, J.I.; Wang, G. Ascorbate induces ten-eleven translocation (Tet) methylcytosine dioxygenase-mediated generation of 5-hydroxymethylcytosine. *J. Biol. Chem.* **2013**, *288*, 13669–13674. [CrossRef] [PubMed]

14. Blaschke, K.; Ebata, K.T.; Karimi, M.M.; Zepeda-Martinez, J.A.; Goyal, P.; Mahapatra, S.; Tam, A.; Laird, D.J.; Rao, A.; Lorincz, M.C.; et al. Vitamin C induces Tet-dependent DNA demethylation and a blastocyst-like state in ES cells. *Nature* **2013**, *500*, 222–226. [CrossRef] [PubMed]

15. Zhang, T.Y.; Meaney, M.J. Epigenetics and the environmental regulation of the genome and its function. *Annu. Rev. Psychol.* **2010**, *61*, 439–466. [CrossRef] [PubMed]

16. Lindblad, M.; Tveden-Nyborg, P.; Lykkesfeldt, J. Regulation of vitamin C homeostasis during deficiency. *Nutrients* **2013**, *5*, 2860–2879. [CrossRef] [PubMed]

17. Hornig, D. Distribution of ascorbic acid, metabolites and analogues in man and animals. *Ann. N. Y. Acad. Sci.* **1975**, *258*, 103–118. [CrossRef] [PubMed]

18. Hughes, R.E.; Hurley, R.J.; Jones, P.R. The retention of ascorbic acid by guinea-pig tissues. *Br. J. Nutr.* **1971**, *6*, 433–438. [CrossRef]

19. Vissers, M.C.; Bozonet, S.M.; Pearson, J.F.; Braithwaite, L.J. Dietary ascorbate intake affects steady state tissue concentrations in vitamin C-deficient mice: Tissue deficiency after suboptimal intake and superior bioavailability from a food source (kiwifruit). *Am. J. Clin. Nutr.* **2011**, *93*, 292–301. [CrossRef] [PubMed]

20. Gariballa, S. Poor vitamin C status is associated with increased depression symptoms following acute illness in older people. *Int. J. Vitam. Nutr. Res.* **2014**, *84*, 12–17. [CrossRef] [PubMed]

21. Cheraskin, E.; Ringsdorf, W.M., Jr.; Medford, F.H. Daily vitamin C consumption and fatigability. *J. Am. Geriatr. Soc.* **1976**, *24*, 136–137. [CrossRef] [PubMed]

22. Prohan, M.; Amani, R.; Nematpour, S.; Jomehzadeh, N.; Haghighizadeh, M.H. Total antioxidant capacity of diet and serum, dietary antioxidant vitamins intake, and serum hs-CRP levels in relation to depression scales in university male students. *Redox Rep.* **2014**, *19*, 133–139. [CrossRef] [PubMed]

23. De Oliveira, I.J.; de Souza, V.V.; Motta, V.; Da-Silva, S.L. Effects of Oral Vitamin C Supplementation on Anxiety in Students: A Double-Blind, Randomized, Placebo-Controlled Trial. *Pak. J. Biol. Sci.* **2015**, *18*, 11–18. [CrossRef] [PubMed]

24. Wang, Y.; Liu, X.J.; Robitaille, L.; Eintracht, S.; MacNamara, E.; Hoffer, L.J. Effects of vitamin C and vitamin D administration on mood and distress in acutely hospitalized patients. *Am. J. Clin. Nutr.* **2013**, *98*, 705–711. [CrossRef] [PubMed]

25. Carr, A.C.; Bozonet, S.M.; Pullar, J.M.; Vissers, M.C. Mood improvement in young adult males following supplementation with gold kiwifruit, a high-vitamin C food. *J. Nutr. Sci.* **2013**, *2*, e24. [CrossRef] [PubMed]

26. Huck, C.J.; Johnston, C.S.; Beezhold, B.L.; Swan, P.D. Vitamin C status and perception of effort during exercise in obese adults adhering to a calorie-reduced diet. *Nutrition* **2013**, *29*, 42–45. [CrossRef] [PubMed]

27. Zhang, M.; Robitaille, L.; Eintracht, S.; Hoffer, L.J. Vitamin C provision improves mood in acutely hospitalized patients. *Nutrition* **2011**, *27*, 530–533. [CrossRef] [PubMed]

28. McNair, D.; MaH, J.W.P. *Profile of Mood States Technical Update*; North Tonawnada: New York, NY, USA, 2005.

29. Carr, A.C.; Pullar, J.M.; Moran, S.; Vissers, M.C. Bioavailability of vitamin C from kiwifruit in non-smoking males: Determination of 'healthy' and 'optimal' intakes. *J. Nutr. Sci.* **2012**, *1*, e14. [CrossRef] [PubMed]

30. Carr, A.C.; Bozonet, S.M.; Pullar, J.M.; Simcock, J.W.; Vissers, M.C. A randomized steady-state bioavailability study of synthetic versus natural (kiwifruit-derived) vitamin C. *Nutrients* **2013**, *5*, 3684–3695. [CrossRef] [PubMed]

31. Lykkesfeldt, J.; Poulsen, H.E. Is vitamin C supplementation beneficial? Lessons learned from randomised controlled trials. *Br. J. Nutr.* **2010**, *103*, 1251–1259. [CrossRef] [PubMed]

32. Bourre, J.M. Effects of nutrients (in food) on the structure and function of the nervous system: Update on dietary requirements for brain. Part 1: Micronutrients. *J. Nutr. Health Aging* **2006**, *10*, 377–385. [PubMed]

33. Gomez-Pinilla, F. Brain foods: The effects of nutrients on brain function. *Nat. Rev. Neurosci.* **2008**, *9*, 568–578. [CrossRef] [PubMed]

34. Kaplan, B.J.; Fisher, J.E.; Crawford, S.G.; Field, C.J.; Kolb, B. Improved mood and behavior during treatment with a mineral-vitamin supplement: An open-label case series of children. *J. Child Adolesc. Psychopharmacol.* **2004**, *14*, 115–122. [CrossRef] [PubMed]

35. Rucklidge, J.J.; Eggleston, M.J.F.; Johnstone, J.M.; Darling, K.; Frampton, C.M. Vitamin-mineral treatment improves aggression and emotional regulation in children with ADHD: A fully blinded, randomized, placebo-controlled trial. *J. Child Psychol. Psychiatry* **2018**, *59*, 232–246. [CrossRef] [PubMed]

36. Rucklidge, J.; Taylor, M.; Whitehead, K. Effect of micronutrients on behavior and mood in adults With ADHD: Evidence from an 8-week open label trial with natural extension. *J. Atten. Disord.* **2011**, *15*, 79–91. [CrossRef] [PubMed]

37. Sarris, J.; Logan, A.C.; Akbaraly, T.N.; Amminger, G.P.; Balanza-Martinez, V.; Freeman, M.P.; Hibbeln, J.; Matsuoka, Y.; Mischoulon, D.; Mizoue, T.; et al. Nutritional medicine as mainstream in psychiatry. *Lancet Psychiatry* **2015**, *2*, 271–274. [CrossRef]

38. Mischoulon, D.; Freeman, M.P. Omega-3 fatty acids in psychiatry. *Psychiatr. Clin. N. Am.* **2013**, *36*, 15–23. [CrossRef] [PubMed]

39. Polak, M.A.; Richardson, A.C.; Flett, J.A.M.; Brookie, K.L.; Conner, T.S. Measuring Mood: Considerations and Innovations for Nutrition Science. In *Nutrition for Brain Health and Cognitive Performance*; Dye, L., Best, T., Eds.; Taylor Japan and Francis: London, UK, 2015; pp. 93–119.

40. Diliberto, E.J.; Daniels, A.J., Jr.; Viveros, O.H. Multicompartmental secretion of ascorbate and its dual role in dopamine beta-hydroxylation. *Am. J. Clin. Nutr.* **1991**, *54*, 1163S–1172S. [CrossRef] [PubMed]

41. Hoehn, S.K.; Kanfer, J.N. Effects of chronic ascorbic acid deficiency on guinea pig lysosomal hydrolase activities. *J. Nutr.* **1980**, *110*, 2085–2094. [CrossRef] [PubMed]

42. Deana, R.; Bharaj, B.S.; Verjee, Z.H.; Galzigna, L. Changes relevant to catecholamine metabolism in liver and brain of ascorbic acid deficient guinea-pigs. *Int. J. Vitam. Nutr. Res.* **1975**, *45*, 175–182. [PubMed]

43. Bornstein, S.R.; Yoshida-Hiroi, M.; Sotiriou, S.; Levine, M.; Hartwig, H.G.; Nussbaum, R.L.; Eisenhofer, G. Impaired adrenal catecholamine system function in mice with deficiency of the ascorbic acid transporter (SVCT2). *FASEB J.* **2003**, *17*, 1928–1930. [CrossRef] [PubMed]

44. May, J.M.; Qu, Z.C.; Meredith, M.E. Mechanisms of ascorbic acid stimulation of norepinephrine synthesis in neuronal cells. *Biochem. Biophys. Res. Commun.* **2012**, *426*, 148–152. [CrossRef] [PubMed]

45. Kuzkaya, N.; Weissmann, N.; Harrison, D.G.; Dikalov, S. Interactions of peroxynitrite, tetrahydrobiopterin, ascorbic acid, and thiols: Implications for uncoupling endothelial nitric-oxide synthase. *J. Biol. Chem.* **2003**, *278*, 22546–22554. [CrossRef] [PubMed]

46. May, J.M. Vitamin C transport and its role in the central nervous system. *Subcell. Biochem.* **2012**, *56*, 85–103. [PubMed]

47. Berk, M.; Williams, L.J.; Jacka, F.N.; O'Neil, A.; Pasco, J.A.; Moylan, S.; Allen, N.B.; Stuart, A.L.; Hayley, A.C.; Byrne, M.L.; et al. So depression is an inflammatory disease, but where does the inflammation come from? *BMC Med.* **2013**, *11*, 200. [CrossRef] [PubMed]

48. Udina, M.; Castellvi, P.; Moreno-Espana, J.; Navines, R.; Valdes, M.; Forns, X.; Langohr, K.; Sola, R.; Vieta, E.; Martin-Santos, R. Interferon-induced depression in chronic hepatitis C: A systematic review and meta-analysis. *J. Clin. Psychiatry* **2012**, *73*, 1128–1138. [CrossRef] [PubMed]

49. Zou, W.; Feng, R.; Yang, Y. Changes in the serum levels of inflammatory cytokines in antidepressant drug-naive patients with major depression. *PLoS ONE* **2018**, *13*, e0197267. [CrossRef] [PubMed]

50. Jeon, S.W.; Kim, Y.K. The role of neuroinflammation and neurovascular dysfunction in major depressive disorder. *J. Inflamm. Res.* **2018**, *11*, 179–192. [CrossRef] [PubMed]
51. Liu, T.; Zhong, S.; Liao, X.; Chen, J.; He, T.; Lai, S.; Jia, Y. A Meta-Analysis of Oxidative Stress Markers in Depression. *PLoS ONE* **2015**, *10*, e0138904. [CrossRef] [PubMed]
52. Carr, A.C.; Rosengrave, P.C.; Bayer, S.; Chambers, S.; Mehrtens, J.; Shaw, G.M. Hypovitaminosis C and vitamin C deficiency in critically ill patients despite recommended enteral and parenteral intakes. *Crit Care* **2017**, *21*, 300. [CrossRef] [PubMed]
53. Carr, A.C.; Maggini, S. Vitamin C and Immune Function. *Nutrients* **2017**, *9*, e1211. [CrossRef] [PubMed]
54. Johnston, C.S.; Solomon, R.E.; Corte, C. Vitamin C status of a campus population: College students get a C minus. *J. Am. Coll. Health* **1998**, *46*, 209–213. [CrossRef] [PubMed]
55. Szczuko, M.; Seidler, T.; Stachowska, E.; Safranow, K.; Olszewska, M.; Jakubowska, K.; Gutowska, I.; Chlubek, D. Influence of daily diet on ascorbic acid supply to students. *Roczniki Państwowego Zakładu Higieny* **2014**, *65*, 213–220. [PubMed]
56. Cahill, L.; Corey, P.N.; El-Sohemy, A. Vitamin C deficiency in a population of young Canadian adults. *Am. J. Epidemiol.* **2009**, *170*, 464–471. [CrossRef] [PubMed]
57. Pullar, J.M.; Bayer, S.; Carr, A.C. Appropriate Handling, Processing and Analysis of Blood Samples Is Essential to Avoid Oxidation of Vitamin C to Dehydroascorbic Acid. *Antioxidants* **2018**, *7*, 29. [CrossRef] [PubMed]

antioxidants

MDPI

Article

Spatial Memory Dysfunction Induced by Vitamin C Deficiency Is Associated with Changes in Monoaminergic Neurotransmitters and Aberrant Synapse Formation

Stine Normann Hansen, Anne Marie V. Schou-Pedersen, Jens Lykkesfeldt and Pernille Tveden-Nyborg *

Section for Experimental Animal Models, Department of Veterinary and Animal Sciences, University of Copenhagen, Thorvaldensvej 57, Ground Floor, 1870 Frederiksberg C, Denmark; snoha@sund.ku.dk (S.N.H.); am.schoupedersen@sund.ku.dk (A.M.V.S.-P.); jopl@sund.ku.dk (J.L.)
* Correspondence: ptn@sund.ku.dk; Tel.: +45-3533-3167

Received: 18 May 2018; Accepted: 27 June 2018; Published: 29 June 2018

Abstract: Vitamin C (vitC) is important in the developing brain, acting both as an essential antioxidant and as co-factor in the synthesis and metabolism of monoaminergic neurotransmitters. In guinea pigs, vitC deficiency results in increased oxidative stress, reduced hippocampal volume and neuronal numbers, and deficits in spatial memory. This study investigated the effects of 8 weeks of either sufficient (923 mg vitC/kg feed) or deficient (100 mg vitC/kg feed) levels of dietary vitC on hippocampal monoaminergic neurotransmitters and markers of synapse formation in young guinea pigs with spatial memory deficits. Western blotting and high performance liquid chromatography (HPLC) were used to quantify the selected markers. VitC deficiency resulted in significantly reduced protein levels of synaptophysin ($p = 0.016$) and a decrease in 5-hydroxyindoleacetic acid/5-hydroxytryptamine ratio ($p = 0.0093$). Protein expression of the N-methyl-D-aspartate receptor subunit 1 and monoamine oxidase A were reduced, albeit not reaching statistical significance ($p = 0.0898$ and $p = 0.067$, respectively). Our findings suggest that vitC deficiency induced spatial memory deficits might be mediated by impairments in neurotransmission and synaptic development.

Keywords: Cavia porcellus; memory deficit; hippocampus; synapse formation; monoaminergic neurotransmitters

1. Introduction

Vitamin C (vitC) deficiency is a surprisingly common nutritional insufficiency affecting around 15% of the Western population [1–3], including subpopulations such as pregnant women and young children [4,5]. The vitamin is a powerful antioxidant, and crucial in the developing brain, where antioxidant defenses are still immature and a high cellular metabolism gives rise to increased levels of reactive oxygen species [6,7]. In the face of dietary depletion, vitC levels in the brain are maintained at approximately 25% of saturated values—as opposed to more extensive reductions in most other organs [8,9], suggesting that the nutrient is of high importance in this tissue. Early life vitC deficiency has been shown to cause impairments in spatial memory, decrease hippocampal volume and neuron numbers in guinea pigs [10,11], a species dependent on dietary vitC akin to humans [12,13]. However, the molecular mechanisms behind the recorded memory deficits are largely undisclosed.

In addition to being pivotal in maintaining brain redox homeostasis, cerebral vitC is linked to glutamatergic neurotransmission and protection against glutamate induced neuronal excitotoxicity [14–16]. VitC is also involved in monoaminergic neurotransmission (dopamine, norepinephrine and possibly serotonin; 5-hydroxytryptamine (5-HT)) [16–18]. Though vitC's specific role in the brain has not been conclusively elaborated, it acts as a co-factor donating electrons to dopamine-β-hydroxylase and is suggested to play a role in monoaminergic neurotransmitter synthesis by keeping the co-factor tetrahydrobiopterin (BH4) in its reduced form [14,16,18,19]. Furthermore, alterations of synaptic structure and associated proteins and receptors have been shown to underlie cognitive dysfunction, behavioral changes and memory formation [20–22]. Thus, either through secondary (antioxidant-mediated) or direct effects on neuronal signaling pathways and dendrite development, vitC deficiency can be speculated to lead to aberrant neurotransmission in the hippocampus, hereby causing the reported spatial memory deficits in young guinea pigs.

In addition, dendrite development is crucial in establishing and maintaining synaptic contacts between neurons [21,23], and decreased dendritic arborization and clustering of the excitatory α-amino-3-hydroxy-5-methyl-4-isoxazolepropionic acid (AMPA) subtype and glutamate receptor subunit (GluR)1 following ablation of neuronal vitC transport has been shown in vitro, suggesting vitC to be a key component for neuronal outgrowth and subsequent signal transduction [14]. We have previously shown impaired hippocampal function measured by decreased spatial memory performance in the Morris Water Maze, in coherence with significantly reduced hippocampal neuron numbers in guinea pigs subjected to vitC deficiency during early life and until reproductive maturity [10]. With off-set in these findings, the current study explores the hypothesis that the recorded memory deficits resulting from vitC deficiency are caused by decreased neuronal signal transmission by reducing monoaminergic neurotransmitters and/or synapse formation and function.

2. Materials and Methods

2.1. Animals

All experiments were approved by the Danish Animal Experiments Inspectorate under the Ministry of Environment and Food (2007/561-1298). The behavioral and histological findings from the in vivo study have previously been published [10].

Briefly, 27, five to six days old female Dunkin Hartley guinea pigs (Charles River Laboratories, Kisslegg, Germany) were weight-stratified and randomly allocated to receive either 923 mg/kg vitC diet by analysis (CTRL, $n = 15$) or 100 mg/kg vitC diet by analysis (DEF, $n = 12$) (Special Diets Services, Dietex International Ltd., Witham, UK) [10]. All animals were inspected and handled daily by trained personnel and allowed ad libitum access to feed, hay (without vitC by analysis) and water. At 52–53 days of age, the animals were subjected to the Morris Water Maze (MWM) test regime, as previously published [10]. At 60–61 days of age, the animals were anesthetized with 0.175 mL/100 g bodyweight Zoletil mix, consisting of 0.465 mg/mL Zoletil-50 (Virbac SA, Carros Cedex, France), 2 mg/mL Xylazin (Narcoxyl, Intervet Int., Boxmeer, The Netherlands), and 1 mg/mL butorphanol (Torbugesic, ScanVet, Fredensborg, Denmark) and briefly supplemented with carbon dioxide inhalation. After disappearance of voluntary reflexes, an intracardiac blood sample was obtained and the animal euthanized by exsanguination [10].

The brain was excised, washed in ice-cold phosphate buffered saline (PBS), and divided into hemispheres. A subset of the hemispheres was randomly allocated to stereological analyses (previously published) [10]. The hippocampus was removed from the remaining left or right hemisphere as determined by randomization, snap frozen in liquid nitrogen and stored at −80 °C until further processing.

2.2. Monoaminergic Neurotransmitters

The analysis of monoaminergic neurotransmitters in the hippocampus was carried out by high performance liquid chromatography (HPLC) as previously described [24]. All samples were analyzed in triplicate and in a randomized order.

2.3. Protein Extraction

The protein was extracted as previously described with some modifications [25]. In short, 40 mg of frozen hippocampal tissue was excised on ice before adding 500 µL RIPA buffer (50 mmol/L tris pH 8.0, 150 mmol/L sodium chloride, 1% Triton X-100, 0.5% sodium deoxycholate and 0.1% sodium dodecyl sulfate) with 1:100 protease inhibitor cocktail (Sigma-Aldrich, Darmstadt, Germany) and 1:100 phosphatase inhibitor cocktail (Sigma-Aldrich, Darmstadt, Germany) and homogenized by mortar and pestle on ice. The samples were centrifuged for 10 min at 12,000 rpm at 4 °C and the supernatant divided in aliquots and stored at −80 °C. In addition, another 10 mg of hippocampal tissue was excised on ice before adding 250 µL of Tissue Protein Extraction Reagent (T-PER) (Thermo Fisher Scientific, Waltham, MA, USA) with 1:100 protease inhibitor cocktail (Sigma-Aldrich, Darmstadt, Germany) and 1:100 phosphatase inhibitor cocktail (Sigma-Aldrich, Darmstadt, Germany). The samples were centrifuged at $10,000\times g$ for five minutes at 4 °C according to manufacturer's instructions and the supernatant divided in aliquots and stored at −80 °C. Two animals from the CTRL group were excluded for technical reasons, leaving the CTRL group size at $n = 13$. Protein concentrations were determined using a commercial BCA kit according to manufacturer's instructions (Merck Millipore, Darmstadt, Germany).

2.4. Western Blotting

The Western blotting procedure was carried out on samples in duplicates and in a randomized order, as previously described [25]. The amount of protein (determined by a dilution series) was adjusted to 11.25 µL with ultrapure water before adding 3.75 µL Laemmli buffer (Hercules, CA, USA) with 1:10 mercaptoethanol (Sigma-Aldrich, Darmstadt, Germany). After denaturing for 10 min at 70 °C, the samples were loaded on a 7.5% Criterion™ TGX™ Precast Midi Protein Gel, 26 well (Bio Rad, Hercules, CA, USA, 15 µL/well) and the electrophoresis was run for approximately 40 min before transferring proteins to a polyvinylidene diflouride (PVDF) membrane. Samples were normalized to total protein levels (REVERT™ Total Protein Stain, Li-Cor, Lincoln, NE, USA). Every blot included positive and negative control samples and a calibrator to account for inter-membrane variation.

The following antibodies were applied: Anti-monoamine oxidase A (MAOA) (ab126751, Abcam, Cambridge, UK, 1:2000, 10 µg protein) with IRDye® 680RD Donkey-anti-Rabbit IgG (Li-Cor, Lincoln, NE, USA; 1:15,000) as the secondary antibody, anti-GluR1 (ab183797, Abcam, Cambridge, UK, 1:1000, 30 µg protein) with IRDye® 680RD Donkey-anti-Rabbit IgG (Li-Cor, Lincoln, NE, USA; 1:15,000) as the secondary antibody, anti-tyrosine hydroxylase (TH) (ab75875, Abcam, Cambridge, UK, 1:500, 40 µg protein) with IRDye® 680RD Donkey-anti-Rabbit IgG (Li-Cor, Lincoln, NE, USA; 1:15,000) as the secondary antibody, anti-tryptophan hydroxylase (TpH) 2 (AV34141, Sigma-Aldrich, Darmstadt, Germany, 1:2000, 20 µg protein) with IRDye® 680RD Donkey-anti-Rabbit IgG (Li-Cor, Lincoln, NE, USA; 1:15,000) as the secondary antibody, anti-post-synaptic-density-protein-95 (PSD-95) (D27E11 3450, Cell Signaling Technology, Boston, MA, USA, 1:1000, 20 µg protein) with IRDye® 680RD Donkey-anti-Rabbit IgG (Li-Cor, Lincoln, NE, USA; 1:15,000) as the secondary antibody, anti-synaptophysin (ab8049, Abcam, Cambridge, UK, 1:1000, 20 µg protein) with IRDye® 680RD Donkey-anti-Mouse IgG (Li-Cor, Lincoln, NE, USA; 1:15,000) as the secondary antibody, anti-neuronal nuclei marker (NeuN) (MAB377, Merck Milipore, Burlington, MA, USA, 1:2000, 20 µg protein) with IRDye® 680RD Donkey-anti-Mouse IgG (Li-Cor, Lincoln, NE, USA; 1:15,000) as the secondary antibdy, anti-glial fibrillary acidic protein (GFAP) (ab7260, Abcam, Cambridge, UK, 1:20,000, 20 µg protein) with IRDye® 800CW Donkey-anti-Rabbit IgG (Li-Cor, Lincoln, NE, USA; 1:15,000) as the secondary antibody and anti-N-methyl-D-aspartate receptor subunit 1 (NMDAR1) (ab77264, Abcam, Cambridge, UK, 1:2000, 20 µg protein) with IRDye® 800CW Donkey-anti-Goat IgG (Li-Cor, Lincoln, NE, USA; 1:15,000) as the secondary antibody. All secondary antibodies were applied for one hour at r/t.

Synaptophysin and NMDAR1 were analyzed using T-PER Tissue Protein Extraction Reagent (Thermo Fisher Scientific, Waltham, MA, USA) extracted protein and the remaining markers using

RIPA extracted protein. The subsequent analyses of staining intensity were completed through Image Studio 5.2 (Li-Cor, Lincoln, NE, USA) by an observer blinded to the experimental groups.

2.5. Statistics

Statistical analyses were performed by GraphPad Prism 7 (GraphPad Software, La Jolla, CA, USA). Student's *t*-test was used for both the neurotransmitter and Western blot analyses. In the event of nonhomogeneous variances, the data was log-transformed or Welch's *t*-test applied. All results are presented as mean $\pm$ SD (standard deviation) or geometric mean (95% confidence interval). A *p*-value < 0.05 was considered statistically significant.

3. Results

The current study is an extension of a previously published in vivo study [10]. In brief, the dietary regimes resulted in ascorbate (Asc; the reduced and active form of vitC) plasma concentrations of 104 ± 34.2 µM in CTRL and 8.5 ± 3.7 µM in DEF, and brain Asc levels of 1256 ± 87.4 nmol/g tissue and 519 ± 99.6 nmol/gram tissue in CTRL and DEF, respectively. Subjected to the Morris Water Maze at day 52–53 of age (around the onset of reproductive maturity), DEF animals displayed significantly impaired performance in the retention test; reduced time spent in platform quadrant, reduced number of crossings of platform area and increased time to first platform area hit ($p < 0.05$, $p < 0.01$ and $p < 0.05$ respectively), compared to CTRL counterparts. A significantly reduced ability to apply a spatial swim pattern was seen in DEF animals compared to CTRL. Stereological quantification of the hippocampus revealed significantly reduced neuron numbers in DEF in all investigated areas (the dentate gyrus, cornu amonis 1 and 2 + 3) [10].

To detect differences in hippocampal neuronal signaling, monoaminergic neurotransmitters and selected metabolites were investigated. The results are shown in Table 1. There was a decreased 5-hydroxyindoleacetic acid (5-HIAA)/5-HT ratio in the DEF group compared with the CTRL ($p = 0.0093$). No other neurotransmitters or their metabolites displayed any differences between the two groups. Dopamine and dopamine metabolites were found to be below detection limit (12 nM for homovanilic acid (HVA), 3.6 nM for 3,4-dihydroxyphenylacetic acid (DOPAC) and 10 nM for dopamine).

Table 1. High performance liquid chromatography (HPLC) detection of monoaminergic neurotransmitters in the hippocampus.

Group/Neurotransmitter	CTRL (*n* = 15)	DEF (*n* = 12)	*p*-Value
MHPG	0.32 ± 0.09	0.35 ± 0.10	NS
Norepinephrine	2.81 ± 1.07	2.32 ± 0.75	NS
MHPG/Norepinephrine *	0.12 (0.10; 0.14)	0.15 (0.12; 0.19)	NS
5-HIAA	0.83 ± 0.16	0.99 ± 0.27	NS
5-HT *	2.07 (1.79; 2.40)	2.2 (1.75; 2.93)	NS
5-HIAA/5-HT	0.47 ± 0.12	0.36 ± 0.08	$p = 0.0093$
HVA	ND	ND	ND
DOPAC	ND	ND	ND
Dopamine	ND	ND	ND

There is a significant decrease in the 5-HIAA/5-HT ratio in the DEF animals. Statistical analysis was performed by Student's *t*-test. Data is displayed as mean $\pm$ SD or mean (95% confidence interval), * log-transformed data. NS: Not significant; ND: Not detectable. MHPG: 3-methoxy-4-hydroxyphenylglycol; 5-HIAA: 5-hydroxyindoleacetic acid; 5-HT: 5-hydroxytryptamine; HVA: Homovanillic acid; DOPAC: 3,4-dihydroxyphenylacetic acid; CTRL: Control animals; DEF: Deficient animals.

To assess whether the imposed state of vitC deficiency in the brain would result in alteration of hippocampal protein expression, Western blot analyses were performed on selected markers, including markers of neuronal maturation (NeuN) and astrocytes (GFAP), as well as more specific markers linked to synapse formation and neurotransmission, such as synaptophysin and NMDAR1. The results from the Western blot analyses are shown in Figures 1–3.

Though the depicted expression patterns may indicate reduced NeuN levels and increased GFAP levels in DEF animals, this could not be confirmed statistically. Hence no differences were observed in overall markers of neuronal and glial cells, NeuN and GFAP, respectively, between the two diet groups (Figure 1).

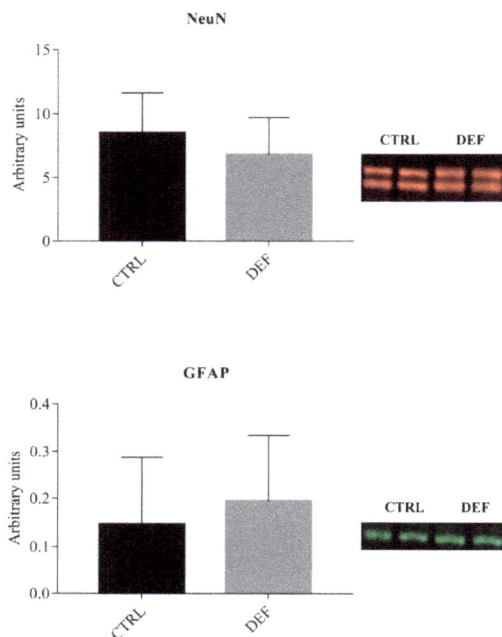

Figure 1. Cellular markers in the hippocampus. The figure depicts the Western blot analyses of the levels of the investigated cellular markers in the hippocampus relative to total protein levels. The two bands in the NeuN samples are consistent with splice variants (confirmed by the manufacturer). No differences between the two groups were detected by Student's *t*-test. Data is shown as mean ± SD, SD: standard deviation. GFAP: Glial-fibrillary-acidic-protein, NeuN: Neuronal nuclei marker. CTRL: Control animals (*n* = 13), DEF: Deficient animals (*n* = 12).

To determine the effects of vitC deficiency on the rate limiting enzymes involved in the metabolism (synthesis and removal) of monoaminergic neurotransmitters, the expression of tyrosine hydroxylase (TH; linked to dopamine and norepinephrine synthesis), tryptophane hydroxylase 2 (TpH2; linked to serotonin synthesis) and monoamine oxidase (MAOA; linked to the removal of dopamine, norepinephrine and serotonin) was assessed. No statistically significant difference between groups could be detected for TH and TpH2. MAOA expression did not reach statistical significance between groups ($p = 0.0844$) (Figure 2).

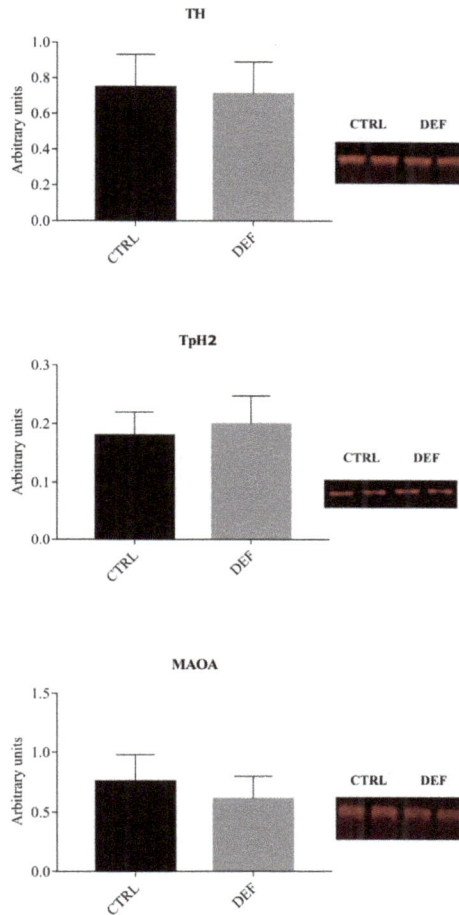

Figure 2. Monoamine synthesizing proteins in the hippocampus. The figure shows the levels of markers associated with monoaminergic neurotransmission in the hippocampus as detected by Western blotting. MAOA may be approaching a decrease in DEF (p = 0.0844), albeit not reaching significance. The expression of the other investigated markers was not different between groups. Data is displayed as mean ± SD and analyzed by Student's t-test. TH: Tyrosine hydroxylase; TpH2: Tryptophan hydroxylase 2; MAOA: Monoamine-oxidase A; CTRL: Control animals (n = 13); DEF: Deficient animals (n = 12).

To evaluate an effect of vitC deficiency on markers of synaptic function (both pre- and postsynaptic), the expression of synaptophysin, NMDAR1, PSD-95 and GluR1 was investigated by Western blotting. A significant decrease in synaptophysin in the DEF group was evident (p = 0.0160), while NMDAR1 approached a decrease in DEF (p = 0.0525) (Figure 3), albeit not reaching statistical significance.

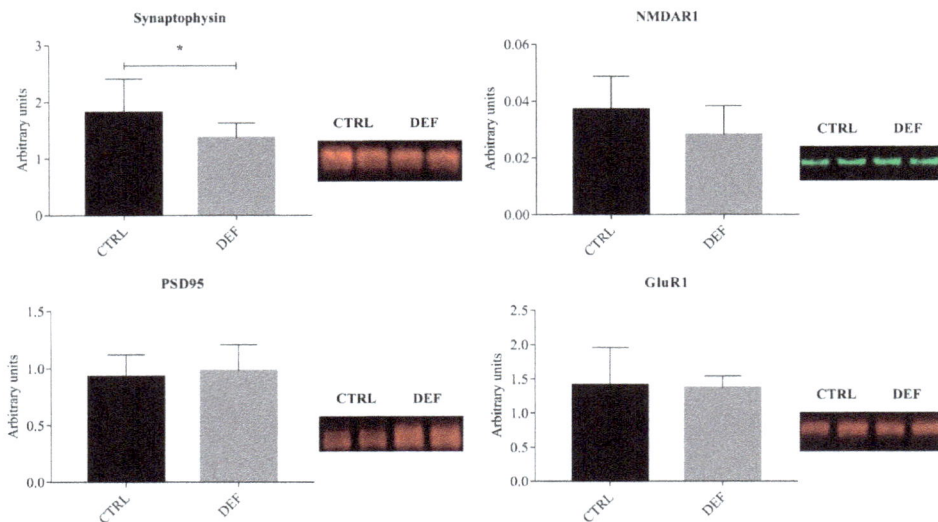

Figure 3. Synapse markers in the hippocampus. The figure shows the results from the Western blotting of markers of synapse formation. Synaptophysin is down-regulated in the DEF group ($p = 0.0160$), while *N*-methyl-D-aspartate receptor 1 (NMDAR1) displays a tendency for down-regulation ($p = 0.0525$). Data is displayed as mean $\pm$ SD and analyzed by Student's or Welch's *t*-test. *: $p < 0.05$; PSD-95: Post-synaptic-density-protein-95; CTRL: Control animals ($n = 13$); DEF: Deficient animals ($n = 12$); GluR1: α-amino-3-hydroxy-5-methyl-4-isoxazolepropionic acid receptor subunit 1.

Expression patterns suggest a NeuN decrease and GFAP increase in DEF, although not substantial enough to reach a significant difference between groups. To assess if a potential difference in neuronal expression was reflected in the expression of additional markers adhering to neuronal function, the expression of synaptic markers NMDAR1, GluR1, synaptophysin and post-synaptic-density-protein-95 relative to neuronal marker NeuN was calculated (relative expression = marker/NeuN). Likewise, the GFAP/NeuN expression ratio was compared to assess glia versus neuron ratio between groups. No statistically significant differences between groups were detected.

4. Discussion

The findings in this study suggest that a vitC deficiency-imposed impairment in spatial memory may be mediated by alterations in monoaminergic neurotransmitter metabolism and aberrant synapse formation in the hippocampus [10].

The decreased 5-HIAA/5-HT ratio in DEF animals supports a putative role of vitC in the metabolism of monoamine neurotransmitters, and that this may subsequently be affected by a state of deficiency, in this case likely due to a decreased 5-HT metabolism. Investigations in vitC deficient mice ($gulo-/-$) found regional alterations (striatum vs. cortex; hippocampus was not analyzed) in the serotonergic system in the brain [17], supporting that there are effects of vitC deficiency, but that these may be presented differently within brain areas.

No apparent difference in hippocampal TpH2 expression—the rate limiting enzyme in 5-HT synthesis [26]—was found between CTRL and DEF groups, in coherence with reports from mice embryos subjected to vitC depletion (SVCT2−/−) or moderate deficiency (SVCT2+/−) [18]. However it may be speculated that the decrease in 5-HT metabolism reflects a compensatory mechanism to keep 5-HT levels intact in response to a low 5-HT synthesis. Indeed, TpH2 is primarily expressed in the raphe nuclei in the brainstem [27,28] from where serotonergic neurons project to the hippocampus [29], hereby rendering TpH2 changes to be undetectable in the hippocampus. Serotonergic neurotransmission in the hippocampus is coupled to several functions including spatial memory [30,31]. An involvement of vitC in serotonergic neurotransmission, and subsequent deviations due to deficiency alongside decreased hippocampal neuron numbers could be a likely cause of the observed spatial memory deficits [10]. Whether additional changes in serotonergic neurotransmission are present—for example, reductions in the 5-HT1a and 5-HT4 receptors [31,32]—remains to be investigated.

Cortical levels of TH protein have been reported to be decreased in Asc depleted mice embryos (SVCT2−/−), but not significantly altered in moderately deficient mice embryos (SVCT2+/−) [18]. This is in agreement with the current finding of no effect on hippocampal TH levels during moderate vitC deficiency, supporting that significant reductions in TH requires extremely low levels of vitC.

The DEF animals displayed a decrease in synapse formation measured as reduced synaptophysin expression, which was positively correlated with Asc levels in the brain, emphasizing a potential direct effect of vitC deficiency on these markers. Synaptophysin is an abundant pre-synaptic vesicle protein involved in regulation of neurotransmitter exo- and endocytosis [33,34] and activity-dependent synapse formation [35]. The synaptophysin knockout mouse model, *Syn*−/−, display decreased learning and memory functions in the object novelty recognition test and MWM [36]. Additionally, several animal models of pathological brain development and memory deficits display decreased levels of synaptophysin [37–39], connecting this marker to the establishment of functional neuronal circuits. DEF animals of the current study displayed significant impairments in the MWM retention test, corresponding well with this finding.

A link between tyrosine phosphorylation of synaptophysin and long term potentiation (LTP)—the cellular hallmark of learning and memory—has been proposed, linking synaptophysin with glutamatergic neurotransmission [36,40,41]. LTP requires the activation of post-synaptic NMDA receptors (though there are also NMDA-independent forms of LTP) and is essential for spatial memory [42–44]. Though not reaching statistical significance, NMDAR1 expression may be decreased by reduced brain Asc status, and reflecting that vitC is important in NMDA receptor-mediated neurotransmission [16,45]. NMDA receptor activation by glutamate gives rise to downstream signaling cascades, increasing AMPA receptor insertion at the post-synaptic site [43]. In the current study, a decrease in the AMPA receptor subunit GluR1 was not detected in DEF, however, in vitro cultures of mouse neurons, devoid of the sodium-dependent vitC transporter 2, display reduced GluR1 clustering [14]. The absence of an effect on the GluR1 subunit in the present study could be due to the moderate state of vitC deficiency and/or represent compensatory mechanisms in vivo. Phosphorylation of GluR1 has been found to be imperative for LTP and memory retention, but not learning, in the MWM [46], suggesting that a decrease in GluR1 phosphorylation may govern the spatial cognitive deficits seen in the vitC deficient animals. Whether this is the case, or if other hippocampal AMPA receptor subunits, such as GluR2, known to be important for spatial memory [47], are affected, requires further investigation.

The DEF animals in this study displayed decreased numbers of neurons in all areas of the hippocampus (the dentate gyrus and cornu ammonis 1 and cornu ammonis 2 and 3) [10]. Thus, a shift in the expression of synaptic markers relative to NeuN expression was explored to clarify any differences which could be masked by an underlying (though on its own un-significant) difference in NeuN expressing cells. The absence of differences in expression patterns relative to NeuN does not support such changes, nor do they support a putative presence of adaptive mechanisms to counteract, for example, decreases in neuronal number and/or a decrease in synapse formation. Furthermore, no alteration in the relative expression of GFAP to NeuN was found in deficient animals, showing that the cellular ratio between neurons and astrocytes was intact in vitC deficient animals. This finding supports that the cellular composition within the hippocampus was not affected by the imposed state of vitC deficiency, however, it is possible that the overall hippocampal volume was reduced in DEF animals hereby giving rise to the recorded reductions in neuronal number. A consistent reduction in overall hippocampal volume due to vitC deficiency in young guinea pigs has previously been reported [11].

Guinea pig specific antibodies of the investigated proteins are not currently commercially available and this represents a serious limitation of the recorded findings due to potential cross-species differences in antibody specificity. Although we included positive controls in all cases, the results of the present study should be verified once antibodies raised against guinea pig target proteins become available.

In conclusion, vitC deficiency-induced spatial memory deficits in young guinea pigs are mediated in part by disturbances in monoaminergic neurotransmission and decreased markers of synapse formation. Further exploration is required to disclose the specific mechanisms by which vitC deficiency affects memory functions in the brain.

Author Contributions: S.N.H., J.L. and P.T.-N. designed the study. A.M.V.S.-P. and S.N.H. conducted the laboratory analysis. All authors contributed to interpretation of the data; S.N.H. and P.T.-N. wrote the draft manuscript. All authors have critically edited and approved the final manuscript.

Funding: This research was funded by the LifePharm Centre for In Vivo Pharmacology and a research grant from The Danish Independent Research Council DFF-4183-00286.

Conflicts of Interest: The authors declare no conflict of interest. The founding sponsors had no role in the design of the study; in the collection, analyses, or interpretation of data; in the writing of the manuscript, and in the decision to publish the results.

References

1. Langlois, K.; Cooper, M.; Colapinto, C.K. Vitamin C status of Canadian adults: Findings from the 2012/2013 Canadian Health Measures Survey. *Health Rep.* **2016**, *27*, 3–10. [PubMed]
2. Hampl, J.S.; Taylor, C.A.; Johnston, C.S. Vitamin C deficiency and depletion in the United States: The Third National Health and Nutrition Examination Survey, 1988 to 1994. *Am. J. Public Health* **2004**, *94*, 870–875. [CrossRef] [PubMed]
3. Schüpbach, R.; Wegmüller, R.; Berguerand, C.; Bui, M.; Herter-Aeberli, I. Micronutrient status and intake in omnivores, vegetarians and vegans in Switzerland. *Eur. J. Nutr.* **2017**, *56*, 283–293. [CrossRef] [PubMed]
4. Ortega, R.M.; Lopez-Sobaler, A.M.; Quintas, M.E.; Martinez, R.M.; Andres, P. The influence of smoking on vitamin C status during the third trimester of pregnancy and on vitamin C levels in maternal milk. *J. Am. Coll. Nutr.* **1998**, *17*, 379–384. [CrossRef] [PubMed]
5. De Oliveira, A.M.; Rondó, P.H.C.; Mastroeni, S.S.; Oliveira, J.M. Plasma concentrations of ascorbic acid in parturients from a hospital in Southeast Brazil. *Clin. Nutr.* **2008**, *27*, 228–232. [CrossRef] [PubMed]
6. Dobbing, J. Later growth of the brain and its vulnerability. *Pediatrics* **1974**, *53*, 2–6. [CrossRef] [PubMed]
7. Ikonomidou, C.; Kaindl, A.M. Neuronal death and oxidative stress in the developing brain. *Antioxid. Redox. Signal.* **2011**, *14*, 1535–1550. [CrossRef] [PubMed]
8. Hughes, R.E.; Hurley, R.J.; Jones, P.R. Retention of ascorbic acid by guinea pig tissues. *Br. J. Nutr.* **1971**, *26*, 433–438. [CrossRef] [PubMed]
9. Hasselholt, S.; Tveden-Nyborg, P.; Lykkesfeldt, J. Distribution of vitamin C is tissue specific with early saturation of the brain and adrenal glands following differential oral dose regimens in guinea pigs. *Br. J. Nutr.* **2015**, *113*, 1539–1549. [CrossRef] [PubMed]
10. Tveden-Nyborg, P.; Johansen, L.K.; Raida, Z.; Villumsen, C.K.; Larsen, J.O.; Lykkesfeldt, J. Vitamin C deficiency in early postnatal life impairs spatial memory and reduces the number of hippocampal neurons in guinea pigs. *Am. J. Clin. Nutr.* **2009**, *90*, 540–546. [CrossRef] [PubMed]
11. Tveden-Nyborg, P.; Vogt, L.; Schjoldager, J.G.; Jeannet, N.; Hasselholt, S.; Paidi, M.D.; Christen, S.; Lykkesfeldt, J. Maternal vitamin C deficiency during pregnancy persistently impairs hippocampal neurogenesis in offspring of guinea pigs. *PLoS ONE* **2012**, *7*, e48488. [CrossRef] [PubMed]
12. Nishikimi, M.; Kawai, T.; Yagi, K. Guinea pigs possess a highly mutated gene for L-gulono-gamma-lactone oxidase, the key ezyme for L-ascorbic-acid biosynthesis missing in this species. *J. Biol. Chem.* **1992**, *267*, 21967–21972. [PubMed]
13. Nishikimi, M.; Fukuyama, R.; Minoshima, S.; Shimizu, N.; Yagi, K. Cloning and chromosomal mapping of the human nonfunctional gene for L-gulono-gamma-lactone oxidase, the ezyme for L-ascorbic-acid biosynthesis missing in man. *J. Biol. Chem.* **1994**, *269*, 13685–13688. [PubMed]

14. Qiu, S.; Li, L.; Weeber, E.J.; May, J.M. Ascorbate transport by primary cultured neurons and its role in neuronal function and protection against excitotoxicity. *J. Neurosci. Res.* **2007**, *85*, 1046–1056. [CrossRef] [PubMed]

15. Sandstrom, M.I.; Rebec, G.V. Extracellular ascorbate modulates glutamate dynamics: Role of behavioral activation. *BMC Neurosci.* **2007**, *8*, 32. [CrossRef] [PubMed]

16. Rebec, G.V.; Pierce, R.C. A vitamin as a neuromodulator—Ascorbate release into the extracellular fluid of the brain regulates dopaminergic and glutaminergic transmission. *Prog. Neurobiol.* **1994**, *43*, 537–565. [CrossRef]

17. Ward, M.S.; Lamb, J.; May, J.M.; Harrison, F.E. Behavioral and monoamine changes following severe vitamin C deficiency. *J. Neurochem.* **2013**, *124*, 363–375. [CrossRef] [PubMed]

18. Meredith, M.E.; May, J.M. Regulation of embryonic neurotransmitter and tyrosine hydroxylase protein levels by ascorbic acid. *Brain Res.* **2013**, *1539*, 7–14. [CrossRef] [PubMed]

19. Diliberto, E.J.; Allen, P.L. Semidehydroascorbate as a product of the enzymic conversion of dopamine to norepinephrine—Coupling of semidehydroascorbate reductase to dopamine-beta-hydroxylase. *Mol. Pharmacol.* **1980**, *17*, 421–426. [PubMed]

20. Beutler, L.R.; Eldred, K.C.; Quintana, A.; Keene, C.D.; Rose, S.E.; Postupna, N.; Montine, T.J.; Palmiter, R.D. Severely impaired learning and altered neuronal morphology in mice lacking NMDA receptors in medium spiny neurons. *PLoS ONE* **2011**, *6*, e28168. [CrossRef] [PubMed]

21. West, A.E.; Greenberg, M.E. Neuronal activity-regulated gene transcription in synapse development and cognitive function. *Cold Spring Harb. Perspect. Biol.* **2011**, *3*. [CrossRef] [PubMed]

22. Rosenberg, T.; Gal-Ben-Ari, S.; Dieterich, D.C.; Kreutz, M.R.; Ziv, N.E.; Gundelfinger, E.D.; Rosenblum, K. The roles of protein expression in synaptic plasticity and memory consolidation. *Front. Mol. Neurosci.* **2014**, *7*, 86. [CrossRef] [PubMed]

23. Copf, T. Impairments in dendrite morphogenesis as etiology for neurodevelopmental disorders and implications for therapeutic treatments. *Neurosci. Biobehav. Rev.* **2016**, *68*, 946–978. [CrossRef] [PubMed]

24. Schou-Pedersen, A.M.V.; Hansen, S.N.; Tveden-Nyborg, P.; Lykkesfeldt, J. Simultaneous quantification of monoamine neurotransmitters and their biogenic metabolites intracellularly and extracellularly in primary neuronal cell cultures and in sub-regions of guinea pig brain. *J. Chromatogr. B* **2016**, *1028*, 222–230. [CrossRef] [PubMed]

25. Søgaard, D.; Lindblad, M.M.; Paidi, M.D.; Hasselholt, S.; Lykkesfeldt, J.; Tveden-Nyborg, P. In vivo vitamin C deficiency in guinea pigs increases ascorbate transporters in liver but not kidney and brain. *Nutr. Res.* **2014**, *34*, 639–645. [CrossRef] [PubMed]

26. Fitzpatrick, P.F. Tetrahydropterin-dependent amino acid hydroxylases. *Annu. Rev. Biochem.* **1999**, *68*, 355–381. [CrossRef] [PubMed]

27. Gutknecht, L.; Kriegebaum, C.; Waider, J.; Schmitt, A.; Lesch, K.P. Spatio-temporal expression of tryptophan hydroxylase isoforms in murine and human brain: Convergent data from Tph2 knockout mice. *Eur. Neuropsychopharmacol.* **2009**, *19*, 266–282. [CrossRef] [PubMed]

28. Zill, P.; Büttner, A.; Eisenmenger, W.; Möller, H.-J.; Ackenheil, M.; Bondy, B. Analysis of tryptophan hydroxylase I and II mRNA expression in the human brain: A post-mortem study. *J. Psychiatr. Res.* **2007**, *41*, 168–173. [CrossRef] [PubMed]

29. Moore, R.Y.; Halaris, A.E. Hippocampal innervation by serotonin neurons of the midbrain raphe in the rat. *J. Comp. Neurol.* **1975**, *164*, 171–183. [CrossRef] [PubMed]

30. Karabeg, M.M.; Grauthoff, S.; Kollert, S.Y.; Weidner, M.; Heiming, R.S.; Jansen, F.; Popp, S.; Kaiser, S.; Lesch, K.-P.; Sachser, N.; et al. 5-HTT deficiency affects neuroplasticity and increases stress sensitivity resulting in altered spatial learning performance in the Morris Water Maze but not in the Barnes Maze. *PLoS ONE* **2013**, *8*, e78238. [CrossRef] [PubMed]

31. Saroja, S.R.; Kim, E.-J.; Shanmugasundaram, B.; Höger, H.; Lubec, G. Hippocampal monoamine receptor complex levels linked to spatial memory decline in the aging C57BL/6J. *Behav. Brain Res.* **2014**, *264*, 1–8. [CrossRef] [PubMed]

32. Hagena, H.; Manahan-Vaughan, D. The serotonergic 5-HT4 receptor: A unique modulator of hippocampal synaptic information processing and cognition. *Neurobiol. Learn. Mem.* **2017**, *138*, 145–153. [CrossRef] [PubMed]

33. Alder, J.; Kanki, H.; Valtorta, F.; Greengard, P.; Poo, M.M. Overexpression of synaptophysin enhances neurotransmitter secretion at Xenopus neuromuscular synapses. *J. Neurosci.* **1995**, *15*, 511–519. [CrossRef] [PubMed]

34. Kwon, S.E.; Chapman, E.R. Synaptophysin regulates the kinetics of synaptic vesicle endocytosis in central neurons. *Neuron* **2011**, *70*, 847–854. [CrossRef] [PubMed]

35. Tarsa, L.; Goda, Y. Synaptophysin regulates activity-dependent synapse formation in cultured hippocampal neurons. *Proc. Natl. Acad. Sci. USA* **2002**, *99*, 1012–1016. [CrossRef] [PubMed]

36. Schmitt, U.; Tanimoto, N.; Seeliger, M.; Schaeffel, F.; Leube, R.E. Detection of behavioral alterations and learning deficits in mice lacking synaptophysin. *Neuroscience* **2009**, *162*, 234–243. [CrossRef] [PubMed]

37. Li, X.; Wang, C.; Wang, W.; Yue, C.; Tang, Y. Neonatal exposure to BDE 209 impaired learning and memory, decreased expression of hippocampal core SNAREs and synaptophysin in adult rats. *Neurotoxicology* **2017**, *59*, 40–48. [CrossRef] [PubMed]

38. Li, Y.; Li, X.; Guo, C.; Li, L.; Wang, Y.; Zhang, Y.; Chen, Y.; Liu, W.; Gao, L. Long-term neurocognitive dysfunction in offspring via NGF/ ERK/CREB signaling pathway caused by ketamine exposure during the second trimester of pregnancy in rats. *Oncotarget* **2017**, *8*, 30956–30970. [CrossRef] [PubMed]

39. Chen, L.-J.; Wang, Y.-J.; Chen, J.-R.; Tseng, G.-F. Hydrocephalus compacted cortex and hippocampus and altered their output neurons in association with spatial learning and memory deficits in rats. *Brain Pathol.* **2017**, *27*, 419–436. [CrossRef] [PubMed]

40. Evans, G.J.O.; Cousin, M.A. Tyrosine phosphorylation of synaptophysin in synaptic vesicle recycling. *Biochem. Soc. Trans.* **2005**, *33*, 1350–1353. [CrossRef] [PubMed]

41. Mullany, P.M.; Lynch, M.A. Evidence for a role for synaptophysin in expression of long-term potentiation in rat dentate gyrus. *NeuroReport* **1998**, *9*, 2489–2494. [CrossRef] [PubMed]

42. Tsien, J.Z.; Huerta, P.T.; Tonegawa, S. The essential role of hippocampal CA1 NMDA receptor–dependent synaptic plasticity in spatial memory. *Cell* **1996**, *87*, 1327–1338. [CrossRef]

43. Herring, B.E.; Nicoll, R.A. Long-term potentiation: From CaMKII to AMPA receptor trafficking. *Annu. Rev. Physiol.* **2016**, *78*, 351–365. [CrossRef] [PubMed]

44. Bliss, T.V.P.; Collingridge, G.L. A synaptic model of memory—Long-term potentiation in the hippocampus. *Nature* **1993**, *361*, 31–39. [CrossRef] [PubMed]

45. Paoletti, P.; Bellone, C.; Zhou, Q. NMDA receptor subunit diversity: Impact on receptor properties, synaptic plasticity and disease. *Nat. Rev. Neurosci.* **2013**, *14*, 383–400. [CrossRef] [PubMed]

46. Lee, H.-K.; Takamiya, K.; Han, J.-S.; Man, H.; Kim, C.-H.; Rumbaugh, G.; Yu, S.; Ding, L.; He, C.; Petralia, R.S.; et al. Phosphorylation of the AMPA receptor GluR1 subunit is required for synaptic plasticity and retention of spatial memory. *Cell* **2003**, *112*, 631–643. [CrossRef]

47. Yan, J.; Zhang, Y.; Jia, Z.; Taverna, F.A.; McDonald, R.J.; Muller, R.U.; Roder, J.C. Place-cell impairment in glutamate receptor 2 mutant mice. *J. Neurosci.* **2002**, *22*, Rc204. [CrossRef] [PubMed]

MDPI

St. Alban-Anlage 66

4052 Basel

Switzerland

Tel. +41 61 683 77 34

Fax +41 61 302 89 18

www.mdpi.com

Antioxidants Editorial Office

E-mail: antioxidants@mdpi.com

www.mdpi.com/journal/antioxidants

www.ingramcontent.com/pod-product-compliance
Lightning Source LLC
Chambersburg PA
CBHW051908210326
41597CB00033B/6066